Charles Dixon

The Nests And Eggs of British Birds

Charles Dixon

The Nests And Eggs of British Birds

ISBN/EAN: 9783744731966

Printed in Europe, USA, Canada, Australia, Japan

Cover: Foto ©Andreas Hilbeck / pixelio.de

More available books at **www.hansebooks.com**

THE NESTS AND EGGS

OF

BRITISH BIRDS

LARGE PAPER EDITION

THE NESTS AND EGGS

OF

BRITISH BIRDS

WHEN AND WHERE TO FIND THEM

BEING

A Handbook to the Oology of the British
Islands

By CHARLES DIXON

AUTHOR OF
'ANNALS OF BIRD-LIFE,' 'IDLE HOURS WITH NATURE,' 'THE BIRDS OF
OUR RAMBLES,' 'THE GAME BIRDS AND WILD FOWL OF THE
BRITISH ISLANDS,' 'THE MIGRATION OF BIRDS,' ETC.

WITH 157 COLOURED ILLUSTRATIONS OF EGGS

LONDON: CHAPMAN AND HALL, LD.
1894

[*All rights reserved*]

RICHARD CLAY & SONS, LIMITED
LONDON & BUNGAY.

PREFACE.

The following pages deal exclusively with the birds that breed within the confines of the British Archipelago; consequently the student will find several familiar species omitted, the Fieldfare and the Redwing for example, they having no claim whatever to be considered in a work which professes to be a Handbook to the Oology of our islands alone.

The idea of a work on the Nests and Eggs of British Birds occurred to me some twelve years ago; and from that time to the present I have been carefully collecting facts, examining specimens, and so on, with the object of forming a comprehensive handbook to British Oology. The results of my studies are now presented to the reader in the following pages. Most of my information has been obtained from personal observation; and with very few exceptions I have taken with my own hands nests and eggs of all our British species; whilst with most of them I have extended my observations over periods of many years.

In the present volume I have endeavoured to foster Oology as a science, not to encourage the indiscriminate collecting of these beautiful objects from the promptings of a mere *bric-à-brac* mania. The nest and eggs of a bird to a great extent reflect the life-history of the bird itself, and vividly illustrate no unimportant part of that bird's economy. If my labours serve to elevate Oology to a higher plane of dignity and importance than that on which it rests at present, it will be a source of unmixed pleasure; whilst if, by setting forth a description of the wonderful, if utilitarian, structures made by our native birds, or the charming and varied characteristics of their beautiful eggs, I can succeed in enlisting a fuller measure of sympathy for the birds themselves, I shall ever view the many years of patient yet loving labour with the highest satisfaction. I commenced to collect nests and eggs when I was barely ten years of age; the passion has grown upon me; and from then to now these wonderful objects have ever won from me an admiration which is only second to that which I have always had for the Birds that make and produce them.

CHARLES DIXON.

February 1893.

CONTENTS.

		PAGE
Raven	*Corvus corax* ...	1
Carrion Crow	,, *corone*	3
Hooded Crow	,, *cornix*	5
Rook	,, *frugilegus*	7
Jackdaw	,, *monedula*	10
Common Chough	*Pyrrhocorax graculus* ...	12
Common Jay	*Garrulus glandarius* ...	15
Magpie	*Pica caudata*	17
Starling	*Sturnus vulgaris*	19
Golden Oriole	*Oriolus galbula*	21
Common Crossbill ...	*Loxia curvirostra*	23
Hawfinch	*Coccothraustes vulgaris* ...	26
Bullfinch	*Pyrrhula vulgaris*	28
House Sparrow	*Passer domesticus*	29
Tree Sparrow	,, *montanus*	32
Greenfinch	*Fringilla chloris*	34
Goldfinch	,, *carduelis*	36
Siskin	,, *spinus*	38
Chaffinch	,, *cœlebs*	40
Linnet	*Linota cannabina*	42
Twite	,, *flavirostris* ...	44
Lesser Redpole	,, *rufescens*	46
Snow Bunting	*Plectrophenax nivalis* ...	48
Reed Bunting	*Emberiza schœniclus* ...	50
Corn Bunting	,, *miliaria* ...	52
Cirl Bunting	,, *cirlus* ...	54
Yellow Bunting	,, *citrinella* ...	55
Sky-Lark	*Alauda arvensis*	58

			PAGE
Wood-Lark ...	...	*Alauda arborea*	60
Pied Wagtail	...	*Motacilla alba yarrellii*	62
White Wagtail ...	...	„ *alba*	64
Gray Wagtail		„ *sulphurea* ...	66
Blue-headed Wagtail		„ *flava*	68
Yellow Wagtail ...		„ *raii*	69
Tree Pipit ...	...	*Anthus trivialis*	71
Meadow Pipit	...	„ *pratensis*	73
Rock Pipit ...	...	„ *obscurus*	75
Common Creeper ...	...	*Certhia familiaris* ...	77
Common Nuthatch ...	...	*Sitta caesia*	79
Bearded Titmouse ...	...	*Panurus biarmicus* ...	81
Long-tailed Titmouse ...	...	*Acredula caudata rosea*	83
Crested Titmouse ...	...	*Parus cristatus*	85
Marsh Titmouse ...	...	*Parus palustris* et *palustris dresseri*	87
Coal Titmouse ...	...	„ *ater* et *ater britannicus*	89
Blue Titmouse ...	...	„ *caeruleus*	91
Great Titmouse ...	...	„ *major*	93
Goldcrest	...	*Regulus cristatus* ...	95
Red-backed Shrike ...	...	*Lanius collurio*	96
Woodchat Shrike ...	...	„ *rufus* ...	99
Chiffchaff	...	*Phylloscopus rufus* ...	100
Willow Wren	...	„ *trochilus* ...	101
Wood Wren	...	„ *sibilatrix* ...	103
Dartford Warbler ...		*Sylvia provincialis* ...	105
Lesser Whitethroat ...	...	„ *curruca* ...	107
Whitethroat	...	„ *cinerea* ...	109
Garden Warbler ...		„ *hortensis* ...	111
Blackcap Warbler ...	...	„ *atricapilla* ...	113
Marsh Warbler ...		*Acrocephalus palustris* ...	115
Reed Warbler ...	...	„ *arundinaceus*	117
Sedge Warbler ...	...	„ *phragmitis* ...	120
Grasshopper Warbler ...		*Locustella locustella* ...	122
Savi's Warbler ...	...	„ *luscinioides* ...	123
Song Thrush ...	...	*Turdus musicus* ...	125
Missel-Thrush ...	..	„ *viscivorus* ...	127
Blackbird ...	...	*Merula merula*	129
Ring Ouzel		„ *torquata*	131

CONTENTS.

		PAGE
Robin	Erithacus rubecula	133
Nightingale	„ luscinia	135
Redstart	Ruticilla phœnicurus	137
Black Redstart	„ tithys	139
Wheatear	Saxicola œnanthe	140
Whinchat	Pratincola rubetra	142
Stonechat	„ rubicola	144
Hedge Accentor	Accentor modularis	146
Dipper	Cinclus aquaticus	148
Common Wren	Troglodytes parvulus	150
St. Kilda Wren	„ parvulus hirtensis	152
Spotted Flycatcher	Muscicapa grisola	155
Pied Flycatcher	„ atricapilla	157
Barn Swallow	Hirundo rustica	159
House Martin	Chelidon urbica	161
Sand Martin	Cotyle riparia	163
Wryneck	Iynx torquilla	164
Green Woodpecker	Gecinus viridis	166
Lesser Spotted Woodpecker	Picus minor	168
Great Spotted Woodpecker	„ major	170
Cuckoo	Cuculus canorus	172
Common Swift	Cypselus apus	175
Common Nightjar	Caprimulgus europæus	177
Hoopoe	Upupa epops	179
Common Kingfisher	Alcedo ispida	181
Barn Owl	Aluco flammeus	182
Wood Owl	Strix aluco	184
Short-eared Owl	Asio brachyotus	185
Long-eared Owl	„ otus	187
Peregrine Falcon	Falco peregrinus	189
Hobby	„ subbuteo	191
Merlin	„ æsalon	193
Kestrel	„ tinnunculus	195
Golden Eagle	Aquila chrysaetus	197
White-tailed Eagle	Haliaëtus albicilla	199
Common Kite	Milvus regalis	201
Honey Buzzard	Pernis apivorus	203
Common Buzzard	Buteo vulgaris	204
Rough-legged Buzzard	Archibuteo lagopus	206

CONTENTS.

		PAGE
Montagu's Harrier	Circus cineraceus	207
Hen Harrier	„ cyaneus	208
Marsh Harrier	„ æruginosus	210
Goshawk	Astur palumbarius	211
Sparrow-Hawk	Accipiter nisus	212
Osprey	Pandion haliaëtus	214
Cormorant	Phalacrocorax carbo	216
Shag	„ graculus	218
Gannet	Sula bassana	220
Mute Swan	Cygnus olor	222
Gray-Lag Goose	Anser cinereus	224
Common Sheldrake	Tadorna cornuta	225
Gadwall	Anas strepera	227
Pintail Duck	„ acuta	229
Wigeon	„ penelope	230
Common Teal	„ crecca	232
Garganey	„ circia	234
Shoveller	„ clypeata	235
Mallard	„ boschas	237
Pochard	Fuligula ferina	239
Tufted Duck	„ cristata	240
Common Scoter	„ nigra	242
Golden-Eye	Clangula glaucion	243
Common Eider	Somateria mollissima	244
Goosander	Mergus merganser	245
Red-breasted Merganser	„ serrator	247
Common Heron	Ardea cinerea	248
Bittern	Botaurus stellaris	250
Stone Curlew	Œdicnemus crepitans	251
Lapwing	Vanellus cristatus	253
Golden Plover	Charadrius pluvialis	255
Dotterel	Eudromias morinellus	257
Kentish Sand Plover	Ægialophilus cantianus	259
Greater Ringed Plover	Ægialitis hiaticula major	260
Oystercatcher	Hæmatopus ostralegus	262
Ruff	Totanus pugnax	264
Common Sandpiper	„ hypoleucus	266
Wood Sandpiper	„ glareola	267
Redshank	„ calidris	269

Greenshank	Totanus glottis	271
Common Curlew	Numenius arquatus	272
Whimbrel	,, phæopus	274
Red-necked Phalarope	Phalaropus hyperboreus	276
Dunlin	Tringa alpina	278
Woodcock	Scolopax rusticola	280
Common Snipe	,, gallinago	282
Richardson's Skua	Stercorarius richardsoni	283
Great Skua	,, catarrhactes	285
Kittiwake	Larus tridactylus	287
Herring Gull	,, argentatus	289
Great Black-backed Gull	,, marinus	291
Lesser Black-backed Gull	,, fuscus	293
Common Gull	,, canus	295
Black-headed Gull	,, ridibundus	297
Sandwich Tern	Sterna cantiaca	300
Roseate Tern	,, dougalli	302
Common Tern	,, hirundo	304
Arctic Tern	,, arctica	306
Lesser Tern	,, minuta	308
Common Guillemot	Uria troile	310
Black Guillemot	,, grylle	312
Razorbill	Alca torda	313
Puffin	Fratercula arctica	315
Leach's Fork-tailed Petrel	Procellaria leachi	317
Stormy Petrel	,, pelagica	319
Fulmar Petrel	Fulmarus glacialis	321
Manx Shearwater	Puffinus anglorum	323
Red-throated Diver	Colymbus septentrionalis	325
Black-throated Diver	,, arcticus	327
Little Grebe	Podiceps minor	329
Great Crested Grebe	,, cristatus	331
Black-necked Grebe	Podiceps nigricollis	333
Sclavonian Grebe	,, cornutus	333
Corn Crake	Crex pratensis	334
Spotted Crake	,, porzana	335
Baillon's Crake	,, bailloni	337
Water-Rail	Rallus aquaticus	338
Waterhen	Gallinula chloropus	340

		PAGE
Common Coot	Fulica atra	... 342
Ring-Dove ...	Columba palumbus ...	... 344
Stock-Dove ...	,, œnas ...	... 346
Rock-Dove ...	,, livia ...	... 348
Turtle-Dove	... Turtur auritus ...	... 350
Pallas's Sand Grouse	... Syrrhaptes paradoxus	... 351
Common Quail ...	... Coturnix communis ...	... 352
Red-legged Partridge	... Caccabis rufa ...	... 354
Common Partridge ...	... Perdix cinerea ...	... 355
Pheasant	... Phasianus colchicus	... 357
Capercaillie	... Tetrao urogallus ...	... 359
Black Grouse	,, tetrix	... 360
Red Grouse ...	... Lagopus scoticus ...	... 362
Ptarmigan ...	... ,, mutus ...	... 364

ADDENDUM.

PAGE 162.—It would appear that the eggs of the House Martin (*Chelidon urbica*) are very exceptionally spotted. I note a recent instance recorded in the *Field*. Without wishing to cast doubt on the *bona fides*, I may remark that no such abnormal variety has ever come under my own observation.

ILLUSTRATIONS.

Plate	I.	...	to face page		1
,,	II.	,,	,,	,,	48
,,	III.	,,	,,	,,	111
,,	IV.	,,	,,	,,	189
	V.	,,	,,	,,	204
,,	VI.	,,		,,	253
,,	VII.	,,		,,	269
,,	VIII.	,,	,,	,,	283
,,	IX.	,,	,,	,,	289
,,	X.	,,	,,	,,	310
,,	XI.	,,	,,	,,	313
,,	XII.	,,	,,	,,	352

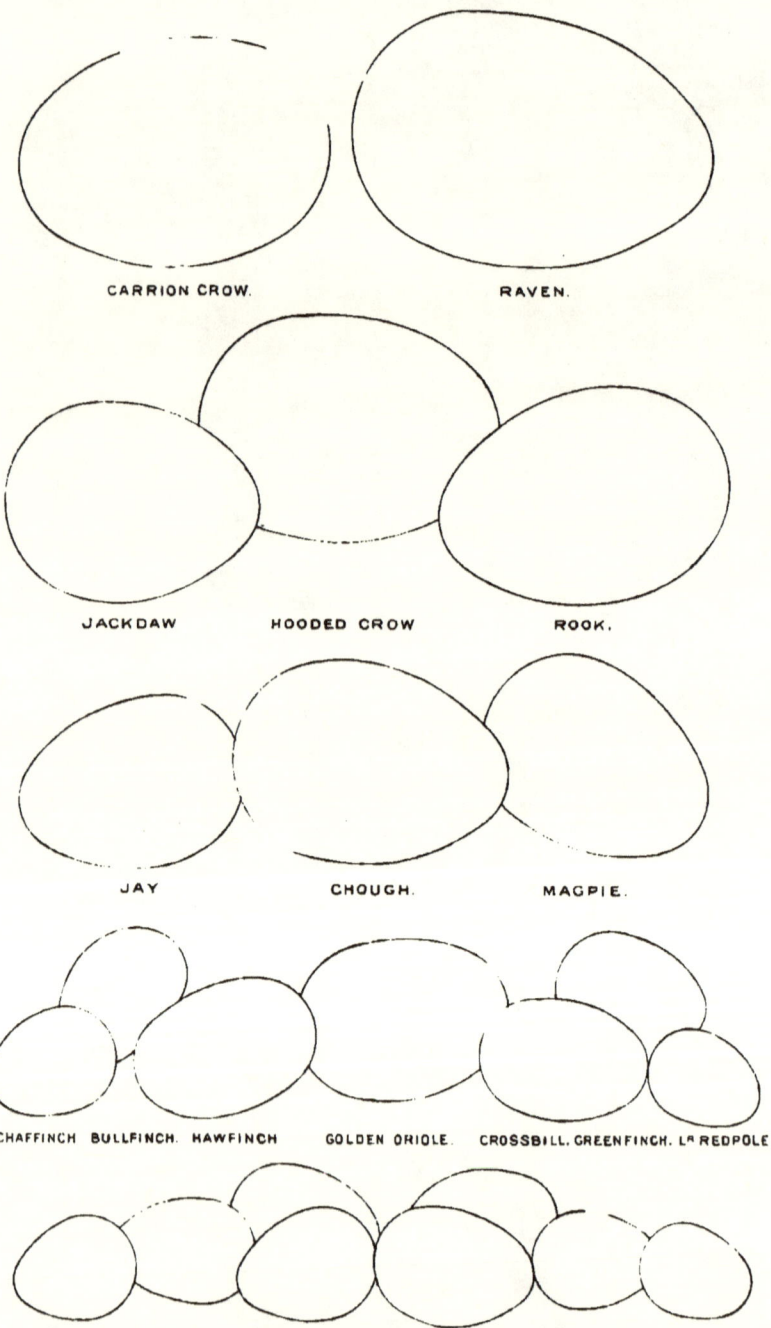

THE NESTS AND EGGS OF BRITISH BIRDS.

Family CORVIDÆ. Genus CORVUS.

RAVEN.

CORVUS CORAX, *Linnæus*.

Single Brooded. Laying season, March and April, sometimes in February.

BRITISH BREEDING AREA: The Raven is slowly vanishing as a breeding species from England. At the present time it breeds here and there on the rocky headlands of the south coast of England, but few are the places where it is allowed to do so in peace. One or two pairs breed regularly on the coast of Devonshire (one pair to my knowledge this season on the rocks near Watcombe). In Wales it is perhaps more common. The great stronghold of this species now is in the wild districts of Ireland and the Highlands, especially in the Hebrides (including St. Kilda), where it nests commonly, even in spite of ceaseless persecution. Inland localities are now nearly deserted, except in very secluded places, and the Raven's great breeding strongholds are on the ocean cliffs.

BREEDING HABITS: The Raven lives in the company

of its mate more or less closely during life, and for years and years will continue to frequent one favourite breeding-place. It is one of the very earliest birds to commence nesting duties, and the eggs are not unfrequently laid before the last of the snow has vanished from its haunts. Formerly the Raven usually nested in a tree; but the incessant persecution that has banished this fine bird from most woodland districts has caused it considerably to alter its domestic arrangements. Most Ravens' nests are now built on lofty and very often inaccessible cliffs. Here a site is found on some ledge, or deep down in a fissure, or even amongst bushes and ivy growing from the rock. The exact spot is often indicated by the white droppings that splash surrounding objects; or the nest itself is generally very large, being often the accumulation of many years, as the birds frequently add to their home each season. The nest is made of sticks and often branches of heather or ling, sometimes with masses of wool clinging to them; turf, roots, moss, wool, fur, and hair form a soft and plentiful lining. It has frequently been remarked that nests of this species built in trees are much more compact than those on cliffs. The parent birds are wary and watchful enough, and are ever ready to attack and beat off any predatory bird that may venture too near their home. During the nesting period the old Ravens may frequently be seen playing and toying in the air above their nesting-place, and uttering by no means unmusical cries.

RANGE OF EGG COLOURATION AND MEASUREMENT: The eggs of the Raven are from three to seven in number; five being an average clutch. They are bluish-green or olive-brown in ground colour, spotted, splashed, and freckled with various shades of olive, and occasionally with smaller markings of very dark brown, nearly black. The underlying spots are paler and not so clearly defined, being often gray or violet in shade. Some eggs are much more sparingly spotted than others. Their shape, too, varies considerably, some being oval, others more rotund, and less frequently, pyriform. Rare varieties are reddish-white in ground colour, spotted with reddish-brown surface markings and violet-gray underlying markings. Average measurement, 2·0 inches in length, by 1·3 inch in breadth. Incubation lasts from nineteen to twenty days, and is performed by both sexes, but the female takes the largest share.

DIAGNOSTIC CHARACTERS : Normally the eggs of the Raven are readily distinguished from those of every other British species by their colour and size ; but in some cases it is impossible to separate them from exceptionally fine eggs of the Carrion Crow, or even the Rook and the Hooded Crow. In such cases it will, however, generally be found that those of the Raven are the *heaviest* for their bulk.

Family CORVIDÆ. Genus CORVUS.

CARRION CROW.

CORVUS CORONE, *Linnæus*.

Single Brooded.   Laying season, April and May.

BRITISH BREEDING AREA : Notwithstanding the war of extermination almost everywhere waged against it, the Carrion Crow is fairly well distributed over the woodland districts and rock-bound coasts of England. In Ireland it is rare and almost replaced by its close ally the Hooded Crow ; the same remarks apply to

Scotland, although it is not so scarce, and certainly breeds sparingly in many districts of the Highlands and is seen in the Hebrides, but I did not observe it in St. Kilda.

BREEDING HABITS: Like the Raven the Carrion Crow pairs for life, and will frequent a certain spot year by year for the purpose of reproduction; but unlike that bird it is one of the latest of the Crow tribe to go to nest. The nests of this species are made in two very distinct situations: viz. on trees in wooded districts, and on cliffs and rocks in hilly and littoral districts. In the woodlands, a tall tree is selected in some secluded spot, often a fir or a pine, frequently an oak, an elm, or a beech, the nest being placed in the topmost branches. When on an inland rock or ocean precipice, precisely similar situations are chosen as those selected by the Raven. The nest is a bulky structure, differing very little in appearance from that of the Rook. As they are often enlarged season by season, some nests are much bigger than others. The nest is made outwardly of large and small sticks, branches of ling, and turf, and lined more or less compactly with roots, wool, moss, fur, and hair, a few feathers, and sometimes quantities of dry withered leaves. The cup containing the eggs is shallow, yet remarkably smooth. Nests in trees are better made than those placed on cliffs, if they are not generally so bulky. The remarkable wariness of this bird is even more intensified during the breeding season. In some districts abroad the Carrion Crow has been known to nest on the ground, and even in this country a large bush has occasionally been selected. Nests of this species are sometimes met with in much-frequented places.

RANGE OF EGG COLOURATION AND MEASUREMENT: The eggs of the Carrion Crow are from three to six in number, five being an average clutch. They are green of various shades in ground colour, spotted, blotched, freckled, and dashed with olive-brown of different tints. Like those of all the other British species in this family they are subject to much variation in the amount and distribution of the markings, which are of two very distinct characters, viz. dark surface spots and paler and grayer underlying ones. Many eggs exhibit a few small dark brown markings, and occasional varieties occur which are almost spotless. Average measurement, 1·7 inch in length, by 1·2 inch in breadth. Incubation lasts from eighteen to twenty days, and is performed by both sexes.

DIAGNOSTIC CHARACTERS: It is impossible to give any reliable character by which the eggs of the Carrion Crow may be distinguished from those of the Hooded Crow or the Rook; but as a rule they are a trifle larger and rounder than the eggs of those species.

Family CORVIDÆ. Genus CORVUS.

HOODED CROW.

CORVUS CORNIX, *Linnæus*.

Single Brooded. Laying season, April and May.

BRITISH BREEDING AREA: Instances of the Hooded Crow breeding in England and Wales are few and irregular, although the bird is said to nest every year in the Isle of Man. In Ireland and Scotland, however, it is widely distributed, especially in the north and west of the latter country, frequenting not only the mainland but most of the islands, from St. Kilda in the west to the Orkneys and Shetlands in the north. In some parts of Scotland it has been known to interbreed with the Carrion Crow, as it habitually does in the valleys of the Elbe and the Yenesay.

BREEDING HABITS: I am of opinion that the Hooded Crow pairs for life, and frequents certain nesting-sites annually wherever it is allowed to remain undisturbed. It is rather a late breeder (although eggs of this species have been recorded exceptionally as early as the middle of March). The Hooded Crow readily adapts itself to circumstances, so far as a nesting-place is concerned. Where trees are scarce it will build on rocks or sea-cliffs, amongst tall ling on the hillsides, or even on the roof of the crofter's hut, as was remarked by Gray. The greater number of nests of this species in our islands are probably built on rocks; but in Siberia, where rocks are scarce and trees plentiful, the latter are most frequently used. Many nests are sometimes built near together, but the Hooded Crow is nothing near so sociable in this respect as the Rook or the Jackdaw. The materials of the nest do not differ from those employed by allied species. Sticks, twigs, the branches of ling and even bones are used to form the outer structure, which is well and warmly lined with moss, wool, hair, fur, feathers, and such-like soft material. Although by no means shy, Hooded Crows are wary enough at the nest, and seem to divine by inspiration whether an intruder can work them harm; being careful to keep out of gunshot.

RANGE OF EGG COLOURATION AND MEASUREMENT: The eggs of the Hooded Crow are from three to six in number, five being an average clutch. They are green, of various shades in ground colour, more or less boldly spotted and blotched with surface markings of olive-brown, and underlying markings of paler brown and gray. They vary considerably in size and shape, an uncommon variety being very pyriform, almost like that of a Snipe. Sometimes they are met with almost spotless. Average measurement, 1·7 inch in length, by 1·2 inch in breadth. Incubation, performed by both sexes, lasts from eighteen to twenty days.

DIAGNOSTIC CHARACTERS: It is impossible to give any reliable character by which the student or collector may be enabled to distinguish the eggs of the Hooded Crow from those of the Carrion Crow and the Rook. The breeding areas of the two latter species are, however, to a certain extent distinct; so that the locality, in the absence of more reliable data, should be of some service in their determination.

Family CORVIDÆ.   Genus CORVUS.

ROOK.

CORVUS FRUGILEGUS, *Linnæus*.

Single Brooded.   Laying season, March and April.

BRITISH BREEDING AREA: The Rook is widely and generally distributed throughout the British Islands, wherever there are trees sufficiently large to support its nest, and the surrounding country is not too barren to furnish it with sustenance. This species is gradually extending its range into many districts, especially in Scotland, where it has followed the planting of trees. More especially has this interesting fact been remarked in the island of Skye, where already there are two extensive colonies, from which pioneers are spreading in various directions with more or less success. As might naturally be inferred, rookeries are more abundant in

England than in other portions of the British Islands. Many rookeries may still be found in towns; and London can boast not a few, although they are gradually decreasing in number and extent.

BREEDING HABITS: The Rook pairs for life. Not only does it continue to frequent the same nesting-sites every season, but it lives in close companionship with its mate throughout the year, and periodically visits its nest trees, as if to assure itself that the home is safe. Tall trees are generally selected for nesting sites—oaks, elms, beeches, horse-chestnuts, planes, and sycamores. Lofty and slender ash and larch trees—mere poles—and in many districts Scotch firs are frequently used. In some rare instances tall holly bushes are selected. As a rule the nests are made in the topmost branches, and very often numbers are built close together, sometimes in large masses. As the Rook is in the habit of adding to its nest each spring, some of the structures are of enormous dimensions, and contain a sackful of sticks or more. In some instances these piles of accumulated sticks have ceased to serve as nests, the owners either being dead or having deserted them. The nest is very firm and compact, made principally of sticks cemented with clay or mud, which latter material usually forms the first or inner lining. This is further lined with turf, moss, wool, dry leaves, bits of straw, and often a few feathers. It is somewhat shallow, but the lining is remarkab smooth and compact. Both male and female assist in building the nest, and as a rule the birds only work at the task in the morning; and in the smaller rookeries never remain all night in the trees until the first eggs are laid. Rooks are remarkably noisy and quarrelsome during the building period, and are very prone to steal materials belonging to their neighbours—a practice which often leads to fatal conflicts. As may often be remarked, many nests are much smaller than others, these being the produce of young and newly-mated birds. Very often an odd nest or so is occasionally seen in trees at some distance from the colony, but these are rarely occupied many seasons in succession; although I have in several instances known them to form the nucleus of a thriving colony. In one case a new rookery was thus formed after a tree had been felled in the main colony. Rooks, however, are most gregarious birds, very conservative, and cling to their favourite breeding-places time out of mind.

RANGE OF EGG COLOURATION AND MEASUREMENT: The eggs of the Rook vary from three to five in number, and should any of the first eggs be taken only the usual clutch seems to be completed, the birds sometimes being content to sit on a single egg. They range from pale blue, nearly white, through various shades of green in ground colour, blotched, spotted, and clouded with olive-brown of varying degrees of intensity and underlying markings of paler brown and gray. Usually the larger end of the shell is most heavily marked. Some specimens are spotted minutely with very dark brown, nearly black; others are streaked with faint olive-green; whilst others, yet again, are almost entirely devoid of colouring matter. They also vary considerably in size and shape, some being very rotund, others elongated. Average measurement, 1·7 inch in length, by 1·2 inch in breadth. Incubation lasts from seventeen to eighteen days, and is performed by both sexes, the female taking the largest share of the task. The eggs are sometimes sat upon as soon as laid.

DIAGNOSTIC CHARACTERS: It is impossible to give any reliable character by which the eggs of the Rook may be distinguished from those of the Carrion and Hooded Crows. The best means of identification is at the nest, the Rooks almost invariably breeding in colonies, the Crows more solitarily. The call-note of the Rook at the nest is nothing near so harsh or guttural as that of the Crows.

Family CORVIDÆ. Genus CORVUS.

JACKDAW.

CORVUS MONEDULA, *Linnæus*.

Single Brooded. Laying season, April and May.

BRITISH BREEDING AREA: The Jackdaw is not only one of the commonest but one of the most widely distributed species in the present family. It breeds more or less abundantly in all suitable districts throughout the British Islands, even extending its range to Skye and the Orkneys, although some localities, apparently suitable in every respect to its requirements, are shunned for no determinable cause. It breeds abundantly on the ocean cliffs, and on many inland rocks, in forest districts, on ruins, castles, and cathedrals, in villages, and even in the busiest of cities.

BREEDING HABITS: Like the Rook, the Jackdaw is not only a life-paired bird, but remarkably gregarious, and breeds in certain places year after year, in most cases using the same nest annually. Like the Carrion Crow, however, it is a somewhat late breeder. The nest is made in a great variety of situations, yet almost invariably in a hole, either of a cliff, a tree, or a building. Instances are on record where the nest is said to have been made amongst ivy growing on cliffs, but such situations must be very exceptional. It is also said occasionally to use a rabbit-burrow for a nesting-place. The Jackdaw breeds in colonies of varying size, according to the relative abundance or scarcity of nesting-sites. Some of these colonies are very extensive, as, for instance, on the cliffs above Devil's Hole in the Peak, and in the hollow trees in Sherwood Forest. Like the Rook, the Jackdaw may be seen almost daily at its nesting-colony all through the year. The nest of this species varies considerably in size. If the selected hole or crevice is large, it is generally filled with materials, and in some cases a vast mass is accumulated, being added to each season; if the site is small but little nest is made, and in many cases no provision whatever is made for the eggs. I have taken the eggs from holes in the trunks of oak trees in Sherwood Forest where the decayed wood alone served for a nest; but on the other hand some of the nests at this colony are several yards in height, the entire hollow trunk or limb having been closely packed with sticks and other rubbish, at the top of which the nest itself has been made. The foundation of the nest is made of sticks and twigs, pieces of turf, and occasionally all kinds of curious litter and rubbish, the cavity containing the eggs being formed of dry grass, pieces of moss, leaves, and straws, and lined with fur, wool, and feathers. Numbers of nests are often made close together, sometimes several in the same hole or cleft. I have known as many as twelve in a single hollow tree. Many nests are quite inaccessible.

RANGE OF EGG COLOURATION AND MEASUREMENT: The eggs of the Jackdaw are from four to six in number, five being the usual clutch. They vary from the palest of blue (almost white) to bluish-green in ground colour, spotted and blotched with very dark brown, nearly black in intensity, olive-brown and grayish-brown, and with underlying markings of pinkish-gray. Some specimens are much more heavily marked than others, and many have the spots small, deep-coloured, and more or less uniformly distributed over the entire surface. Some examples are very sparingly marked, and mostly with underlying spots. In shape and size also they vary considerably. Average measurement, 1·45 inch in length, by 1·0 inch in breadth. Incubation, performed by both sexes, lasts from seventeen to eighteen days.

DIAGNOSTIC CHARACTERS: The clear blue or pale green ground colour, and comparatively few and well-defined markings, readily distinguish the eggs of the Jackdaw from allied British species. The eggs most likely to be confused with them are those of the Chough and certain pale varieties of those of the Magpie, but the clear definition of the surface-spots is a pretty safe guide. It must also be borne in mind that the Jackdaw rarely or never breeds near the colonies of the Choughs and the exposed, domed nest of the Magpie in the branches is ever an unfailing means of distinction.

Family CORVIDÆ.   Genus PYRRHOCORAX.

COMMON CHOUGH.

PYRRHOCORAX GRACULUS (*Linnæus*).

Single Brooded. Laying season, May.

BRITISH BREEDING AREA: The Chough is now one of the rarest and most local of our indigenous birds, and though formerly fairly common in many inland districst, is now almost without exception only met with during the breeding season on the wildest and most inaccessible ocean cliffs. Even here many of its scattered colonies have been deserted, for no apparent cause, within comparatively recent times. Owing to the character of the coast, the breeding-places of the Chough are on the southern and western shores of the British Islands—districts in which the cliffs are lofty and far removed from the busy haunts of men. It breeds locally and sparingly from Dorset, west to Devon and Cornwall. A few pairs still breed on Lundy Island. Scattered colonies are established here and there on the wild rocky coasts of Wales, as well as in one or two localities inland in that portion of our islands; whilst in the Isle of Man a few pairs still continue to breed. It is said also occasionally to breed in Cumberland. On the west coast of Scotland it is fairly plentiful, on the islands of Islay, Jura, and Skye, especially the former, and also in one or two localities on the cliffs of the mainland. In Ireland its chief strongholds are on the coasts of Kerry, Mayo, Donegal and Antrim, Waterford and Cork.

BREEDING HABITS : There is much similarity between the habits of the Chough and the Jackdaw during the season of reproduction. Both birds are gregarious and breed in colonies, both birds are life-paired species, and both breed in holes or covered sites, and continue year after year to rear their young in certain favoured spots, tenanting the same nests each recurring season. On the cliffs the nests are made in the clefts or fissures, or in holes in the roof or sides of caves, often so deep down in the rock as to be absolutely inaccessible; inland a hole in a ruin is sometimes selected. The nest, like that of the Jackdaw, varies a good deal in size, according to the accommodation afforded by the selected hole or fissure, and is composed of sticks, dead branches of heather, and the dry stalks of plants. The cavity containing the eggs is formed of dry grass, roots, wool, fur, and occasionally hair. If the colony is disturbed, the birds fly out of their nest-holes and act in a very Jackdaw-like manner, uttering their noisy and distinctive cries. One cavity sometimes contains several nests.

RANGE OF EGG COLOURATION AND MEASUREMENT: The eggs of the Chough vary from four to six in number, and range in ground colour from white, with a faint tinge of blue or green, to creamy-white, blotched and spotted with various shades of brown and gray, and with underlying markings of violet-gray. Sometimes a few dark brown streaks or scratches occur, usually on the larger end. In the size and distribution of the markings they vary considerably, some having them large and bold, and arranged in a mass or zone round the largest portion of the shell; others are more regularly spotted over the entire surface, and in this type the colour of the marks is paler. Another type has few markings, but bold and large, and scattered here and there over the entire surface. Average measurement, 1·5 inch in length, by 1·1 inch in breadth. Incubation, performed by both parents, lasts from seventeen to eighteen days.

DIAGNOSTIC CHARACTERS: The pale ground colour and comparative indistinct definition of the markings are the most important points of distinction characterizing the eggs of the present species. They are most likely to be confused with eggs of the Jackdaw; but the two species do not breed in company.

Family CORVIDÆ. Genus GARRULUS.

COMMON JAY.
GARRULUS GLANDARIUS (*Linnæus*).

Single Brooded. Laying season, April and May.

BRITISH BREEDING AREA: The Jay, in spite of constant persecution, breeds more or less sparingly throughout the woodland districts of England and Wales. In many places where game is not very strictly preserved, and feathered marauders allowed to dwell in peace, the Jay is a common bird. In Scotland it is much more local and slowly becoming scarce, from the same causes, although its range has been extended with the planting and growth of trees. Its principal quarters in Scotland are the central counties, as far north as Inverness-shire; it does not, however, breed in the Hebrides, and is only a straggler to the Shetlands. In Ireland, where it has also decreased in numbers, its chief breeding area is now in the south-east, in the area confined by the rivers Barrow and Suir.

BREEDING HABITS: The principal breeding-grounds of the Jay are the game coverts, woods, and plantations where the underwood is dense and leafy. The bird is particularly fond of covers where evergreens are plentiful, and is thus specially addicted to shrubberies. In all cases, however, the Jay will nest most abundantly where the cover is thickest. In my opinion this handsome bird is a life-paired species, and may be seen in company with its mate all through the year. As the breeding season approaches the Jay becomes much less noisy and even more skulking in its movements, so that it often safely rears a brood in a cover where its presence has never been suspected. The site for the nest is seldom at any great altitude from the ground, the cover usually not being sufficiently dense above the growth of tangled underwood. A situation is generally selected in a tall bush or sapling, especially in a holly, yew, or other evergreen, and less frequently amongst a mass of woodbine or ivy. I have known the nest to be made in a bunch of twigs and fine branches growing from a tree trunk where a branch has been lopped off. The nest is cup-shaped, and made externally of fine sticks and twigs, occasionally cemented with mud, and thickly lined with fibrous roots right up to the margin. Although made of coarse materials it is very neatly finished. The Jay is non-gregarious during the breeding season.

RANGE OF EGG COLOURATION AND MEASUREMENT: The eggs of the Jay are from five to seven in number. They are grayish-green or pale bluish-green in ground colour, mottled and speckled over the entire surface with olive-brown, and occasionally streaked with a few scratches or irregular lines of dark liver-brown. In many examples the mottled spots become most numerous and confluent towards the larger end, where they form a distinct zone. The eggs of the Jay do not present much variation, although it is not unusual to find an egg paler than the rest in a clutch. Average measurement, 1·25 inch in length, by ·9 inch in breadth. Both parents assist in the task of incubation, which extends over a period of eighteen days.

DIAGNOSTIC CHARACTERS: The small size, indistinct character of the markings, and grayish-green appearance, readily distinguish the eggs of the Jay from all the allied British species.

Family CORVIDÆ. Genus PICA.

MAGPIE.

PICA CAUDATA, *Gerini.*

Single Brooded. Laying season, March and April.

BRITISH BREEDING AREA: The Magpie breeds more or less commonly throughout the wooded districts of the British Islands, and in many places continues steadily to increase in spite of incessant persecution. There can be little doubt that this is due to the bird's great fecundity, and its habit of nesting in tall trees. The Magpie does not breed on the Outer Hebrides, nor does it visit the Orkneys and the Shetlands, but it nests sparingly on the Channel Islands. In Ireland the Magpie is much more abundant than formerly, and its numbers are still perceptibly increasing in many localities.

BREEDING HABITS: Although the great breeding-grounds of the Magpie are situated in or near woods, game coverts, and plantations, a considerable number of nests are made in more open situations, where timber is scarce. This species also pairs for life, and either tenants the same nest yearly, in spite of continued robbery and disturbance, or builds a new structure in the immediate neighbourhood. The nest may be found in almost every kind of forest tree, often in tall bushes or isolated trees in the fields, or even in a low hedgerow. The altitude varies equally as much, from the slender topmost twigs of some woodland giant, to the tangled thicket or hedgerow, not more than six or eight feet from the ground. Another remarkable trait in the Magpie's character is that of nesting close to habitations. Although a remarkably shy and wary bird all the rest of the year, in the breeding season it often becomes most trustful. The site for the nest, when in a tree, is usually amongst the slender branches, sometimes near the extremity of a massive limb, but in bushes or hedges the densest part is chosen. The nest is a large and bulky structure, and when finished completely covered with a dome or roof. The outer part of the nest is first formed of sticks, which are cemented together with large quantities of clay or mud, a lining of this material eventually being made. Then the huge dome is built over, dead thorns being favourite material, a hole being left on the side, near the top or rim of the nest cavity, for ingress. Very often at this stage the nest is left for a day or so to dry, before the copious lining of fibrous roots is added. It has been recently stated that dry grass is also used, but this I deny, at any rate so far as British Magpies are concerned. The whole structure when completed and thoroughly dry is remarkably compact. Occasionally several Magpies' nests may be found at no great distance from each other, but the bird is neither gregarious nor social during the breeding season. It is a wary bird at the nest, slipping off very quietly if disturbed during incubation, but becoming much more bold and demonstrative when the young are hatched.

RANGE OF EGG COLOURATION AND MEASUREMENT: The eggs of the Magpie are from six to nine in number; seven and eight are frequently found, but perhaps six forms the most usual clutch. They range from creamy-white to bluish-green in ground colour, thickly freckled, blotched, and spotted over the entire surface with olive-brown, and occasionally streaked with a few lines of very dark brown. They vary considerably both in shape, in size, and in colouration. Some eggs are very pyriform; others are almost round. Some have little surface-colour upon them, being almost spotless pale blue, with perhaps a few violet-gray underlying markings;

others are grayish-white, thickly freckled with pale ash-brown. Average measurement, 1·35 inch in length, by 1·0 inch in breadth. Incubation, performed chiefly by the female, lasts eighteen days.

DIAGNOSTIC CHARACTERS: The size, abundance, and smallness of the uniformly distributed markings and their brown colour readily distinguish the eggs of the Magpie from all allied species breeding within the British area.

Family STURNIDÆ.   Genus STURNUS.

STARLING.

STURNUS VULGARIS, *Linnæus*.

Double Brooded.   Laying season, April to June.

BRITISH BREEDING AREA: Widely distributed, and breeds more or less abundantly throughout England and Wales, becoming the least abundant perhaps in the more thinly-populated districts of Wales. In Scotland it has increased in numbers, and extended its range to such a very remarkable extent during the past half-century, that it may now be said to breed in every suitable locality. It also breeds throughout the Hebrides (including St. Kilda), and in the Orkneys and Shetlands. In Ireland it becomes much more local during the breeding season, being there most widely distributed and most abundant, especially in the south, during winter.

BREEDING HABITS: The Starling is another life-paired species, and returns annually to its old nesting-place; indeed, like the Rook, it may, in a great many instances, be seen to visit its nesting-site from time to time right through the non-breeding season. Like the Rook and the House Sparrow, it evinces a strong partiality for the dwelling-places of man, and like the latter bird it also displays great aptitude for adapting itself to a variety of conditions during the season of reproduction. The Starling will make its nest almost anywhere, provided a site can be found well protected from the external air. Holes in buildings are now the favourite situations, but great numbers of nests are made in holes in trees and cliffs, in peat-stacks, and less frequently in rabbit-burrows. On St. Kilda I was both surprised and delighted to find my favourite bird actually nesting in holes in the ground, on the bare hillsides, and amongst the rough walls of the "cleats." The Starling is thoroughly gregarious all the year round, and numbers breed in close companionship, the size of the colony depending a good deal on the amount of the accommodation to be had. Instances are on record where the Starling has been known to breed in a Magpie's nest, and to rear its young in an open nest in a tree; but from my lifelong acquaintance with this species I think such exceptions (if true) must be excessively rare. The size of the hole varies a good deal, hence the nest is bulky or small accordingly. It is a rude, slovenly structure, made of straw, grass, and roots, and sometimes lined with a few feathers; rags, twine, or paper will even be used occasionally. Very often a few straws are allowed to dangle out of the entrance, proclaiming the nest to every passer-by. If the eggs are removed from day to day the hen may be encouraged to lay an indefinite number of eggs. I once took no less than forty eggs from one hole during a single season. Odd Starlings' eggs are frequently found in the fields.

RANGE OF EGG COLOURATION AND MEASUREMENT: The eggs of the Starling are from four to six or even seven in number, five being an average clutch. They are pale greenish-blue, somewhat elongated, rather rough in texture, but with considerable gloss, and are spotless. The shade of colour varies a good deal, some eggs being almost white, others much more intense. Exposure to the light soon robs them of much of their colour. Average measurement, 1·2 inch in length, by ·85 inch in breadth. Incubation, performed chiefly by the female, lasts fourteen days.

DIAGNOSTIC CHARACTERS: The eggs of no other species breeding in the British Islands can easily be confused with those of the Starling. Their large size and uniform greenish-blue tinge readily distinguish them from the eggs of all our smaller species.

Family ORIOLIDÆ.   Genus ORIOLUS.

GOLDEN ORIOLE.

ORIOLUS GALBULA, *Linnæus*.

Single Brooded. Laying season, May and June.

BRITISH BREEDING AREA: The Golden Oriole is another melancholy instance of the senseless persecution of rare birds in the British Islands. There is nothing, so far as we can determine, to prevent this handsome and melodious songster from becoming as common on this side of the Channel as it is on the other, if the individuals of this species that almost yearly visit our southern and south-western counties were allowed to live and rear their broods in peace. There can be little or no doubt that the Golden Oriole has bred in Kent, Surrey, Essex, Northamptonshire, and Norfolk, but, alas! the gaudy dress of the male bird is a fatal attraction, and the gunner or greedy collector soon put an end to the naturalist's hope of seeing this handsome species increase and multiply amongst us.

BREEDING HABITS: Like most birds of conspicuous and gaudy plumage, the Golden Oriole is fond of the cover, and, although by no means a shy bird, is careful to conceal itself amongst the leaves when menaced by danger. In Europe the favourite breeding-grounds of the Golden Oriole are groves, the borders of woods, plantations, and well-timbered fields; but in Algeria, where I saw much of this species, the oases, public gardens, and groves of evergreen oaks in the park-like country of the Aures Mountains were the favourite haunts. It is not improbable that this bird pairs for life, although I do not think the same nest is used two seasons in succession. The Golden Oriole very frequently rears its young quite close to houses, and even in towns, in avenues, and pleasure-grounds. The site for the nest is usually in an oak, plane, or fir tree, at distances varying from twenty to forty feet from the ground. The nest, so far as British, or even European birds are concerned, is unique, and cannot possibly be mistaken for that of any other bird. It is invariably suspended, cradle-wise, between a forked horizontal branch, the external materials being deftly woven round the supporting twigs. The outside is made of broad grass, sedge, and strips of bark, amongst which a few dead leaves or even scraps of paper are interwoven; the inside is lined with the fine round flower-stems of grass. During the nesting period the old Orioles are careful not to betray their secret, and are ever ready to drive off any intruding birds. The very characteristic song of the male often proves a guide to the whereabouts of the nest.

RANGE OF EGG COLOURATION AND MEASUREMENT: The eggs of the Golden Oriole are from four to six in number. They are pure white, or white with a yellowish tinge, in ground colour, spotted and speckled with purplish-brown, and generally with a few small pale gray underlying markings. The surface of the shell, though somewhat rough in texture, is polished and glossy. As a rule, the eggs of this species do not exhibit much variety. Average measurement, 1·2 inch in length, by ·9 inch in breadth. Incubation, performed chiefly by the female, lasts from fourteen to fifteen days.

DIAGNOSTIC CHARACTERS: The size, white ground colour, and reddish-purple spots, readily distinguish the eggs of the Golden Oriole from those of every other British species. The unique character of the nest is also another unfailing guide to the identification of the eggs.

Family FRINGILLIDÆ.
Sub-family FRINGILLINÆ.

Genus LOXIA.

COMMON CROSSBILL.

LOXIA CURVIROSTRA, *Linnæus*.

Single Brooded. Laying season, February to April.

BRITISH BREEDING AREA: The Common Crossbill is one of the most local of the species that breed within the limits of the British Islands. In England it breeds locally and irregularly in many of the counties lying south of a line drawn from the Wash to the Bristol Channel; whilst north of this limit it is known to do so in Yorkshire, Durham, and Northumberland. In Scotland it breeds more freely, especially in the central counties; whilst in Ireland it is said by Mr. Ussher to have increased of late years as a breeding species, although its distribution during the nesting season in the sister isle is very imperfectly known. It appears to nest in some of the extreme eastern counties from Down to Waterford and Tipperary. Owing to the excessive shyness of this species in the nesting season, its preference for dense conifer plantations, and the early date of laying, there can be little doubt that the Crossbill is much overlooked.

BREEDING HABITS: The favourite breeding-grounds of the Crossbill are the conifer plantations, the belts and enclosures of Scotch fir, spruce, and larch. It is not improbable that this species pairs for life, but owing to its nomadic habits neither the old nest nor the old locality seem to be visited each season for breeding purposes. The Crossbill, however, is a remarkably social bird, and not only spends the winter in parties in wandering about in quest of food, but during the breeding season lives frequently in companies, and several nests may often be found within a comparatively small area. The nest is generally built amongst the foliage of the Scotch fir, a site being selected at various heights from the ground, sometimes as many as forty or fifty feet, at others not more than four or five feet. Nests are far less frequently found in deciduous trees, owing to the fact that the foliage is not out at the time the Crossbills begin to build. Sometimes the nest is made at a distance from the trunk on a flat branch; at others it is wedged into a fork near the top of the tree. The nest itself very closely resembles one type of that of the Greenfinch, being formed outwardly of twigs loosely twined together, rootlets, and dry grass, and inwardly of wool, fur, and a few feathers and hairs. Sometimes a little moss, lichen, and a few bark strips are employed. The female sits very closely, often allowing herself to be closely scrutinized, or almost touched by the hand, before she slips off her nest. Mr. Norgate remarks of a nest taken by him in Norfolk five years ago: "On March 26th I took a nest of four Crossbills' eggs from a Scotch fir; the hen bird objected to leave the nest even after it was brought down from the tree, when three or four other Crossbills came and fluttered about close to our heads, uttering their peculiar cry and showing their hooked beaks." When disturbed from her eggs the female is often joined by the male, both birds fluttering about in an anxious, restless manner.

RANGE OF EGG COLOURATION AND MEASUREMENT: The eggs of the Crossbill are usually four in number, occasionally three, more exceptionally five. They vary from white to white tinged with green in ground colour, spotted with reddish-brown, and with underlying markings of paler brown. The spots, never very large, are mostly distributed over the larger end of the egg, where they not unfrequently form an irregular zone. Many of the spots often take the form of streaks, and then the colour is exceptionally dark. Average measurement, ·9 inch in length, by ·7 inch in breadth. Incubation, performed by the female, lasts fourteen days.

DIAGNOSTIC CHARACTERS: It is impossible to give any absolutely reliable character by which the eggs of the Crossbill may be distinguished from those of the Greenfinch. The range of colouration is practically the same in each species, but as a rule the eggs of the latter bird are smaller. The date of laying is one reliable characteristic, the Crossbill producing eggs as a rule from one to two months earlier than the Greenfinch.

Family FRINGILLIDÆ. Genus COCCOTHRAUSTES.
Sub-family FRINGILLINÆ.

HAWFINCH.

COCCOTHRAUSTES VULGARIS, *Pallas*.

Single Brooded.   Laying season, May.

BRITISH BREEDING AREA: Owing to the excessive shyness and skulking habits of the Hawfinch, it is somewhat difficult to define its exact distribution during the breeding season. It is decidedly an English bird, not breeding in any other portion of the United Kingdom, although some writers state that it does so in Ireland. Some authorities assert that the Hawfinch has steadily increased in numbers during the past half-century, but I am inclined to attribute this apparent fact to the closer scrutiny and greater number of observers. The Hawfinch breeds locally, and more or less frequently in almost every county of England; most abundant in the home counties, and least so in the extreme north and west.

BREEDING HABITS: For a month or more before the nest is commenced, the parties of Hawfinches that have been leading a more or less nomadic life during the winter separate into pairs and betake themselves to the accustomed nesting-places. Orchards, small woods, fir plantations, and, less frequently, shrubberies are the favourite breeding-grounds of the Hawfinch; but gardens, tall hedges in well-timbered fields, and ivy-clad trees in lanes are also selected. As the breeding season approaches, the birds become even more shy and seclusive in their habits, and the peculiar and characteristic song of the males is almost the only sign of their presence. The nest is usually commenced towards the end of April. A site is selected in the branches of an old lichen-draped fruit tree, or hawthorn, in the pollard top of a hornbeam, amongst ivy, or in the dense branches of a yew, a fir, or a holly. The altitude varies considerably. Some nests are built as much as forty or fifty feet from the ground; others only a few feet. In many cases several pairs make their nest in the same plantation or enclosure, especially in districts where suitable cover is scarce. The nest is made externally of twigs, roots, scraps of lichen, and the dry stalks of various plants, and internally of dry grass, finer roots, and hair. It is large, flat in appearance, and although rudely fabricated outside is neat and well finished within. The hen-bird is a close and silent sitter, and when flushed glides very quietly from her eggs into the surrounding cover. If the first nest be destroyed another attempt to rear a brood is generally made. The Hawfinch becomes even much more silent than usual as soon as the eggs are laid.

RANGE OF EGG COLOURATION AND MEASUREMENT: The eggs of the Hawfinch are from four to six in number, the latter being perhaps more frequent than the former. They vary from pale olive or pale bluish-green to pale brownish-buff in ground colour, streaked and more sparingly spotted with dark olive-brown and pale grayish-brown, becoming almost violet-gray in buff ground-coloured examples. The streaks are frequently intricate, and as pronounced as those on a typical Bunting's egg. The amount of markings varies considerably, and, as a rule, on the eggs on which the spots are largest and most clearly defined the streaks are finer, more scratchy, and paler. Average measurement, ·95 inch in length, by ·75 inch in breadth. Incubation, performed by the female, lasts from fourteen to fifteen days.

DIAGNOSTIC CHARACTERS: The size, combined with the streaky markings and their colour, serve to distinguish the eggs of the Hawfinch from those of every other allied British species.

Family FRINGILLIDÆ.　　　　　　　　Genus PYRRHULA.
Sub-family FRINGILLINÆ.

BULLFINCH.

PYRRHULA VULGARIS, *Temminck.*

Double Brooded.　Laying season, April to June, and even July.

BRITISH BREEDING AREA: Breeds more or less abundantly in all suitable localities throughout the British mainland, becoming rarer and more local in Scotland and Ireland. There can be no doubt whatever that the bird-catcher has well-nigh exterminated this species from many districts where it formerly bred in considerable numbers.

BREEDING HABITS: The Bullfinch mates for life, and all through the year may be seen in pairs, although the old nest is not used season by season, neither, in many cases, is the same locality selected. Like many other birds that lead a nomadic life during winter, the Bullfinch appears to breed in any suitable district it may chance to be in when the time for that event arrives. Like the preceding species the Bullfinch becomes very silent, shy, and retiring in its habits during the season of reproduction, and the love-song of the male generally ceases as soon as the eggs are laid. The chief breeding-haunts of the Bullfinch are plantations of firs, shrubberies, orchards, and dense hedgerows and thickets. The nest is built usually from six to ten or fifteen feet from the ground, in a fork of the branches, or on a flat branch at some distance from the trunk. Externally it is composed of fine twigs intricately interlaced, forming a flat structure, in the centre of which the cup for the eggs is made of roots and hair, and occasionally one or two feathers, or a scrap of wool. During the period of incubation the Bullfinch becomes very quiet and shy, and the female, as is usual with so many Finches, is a close sitter, reluctant to leave her charge, yet slipping very stealthily away when disturbed. The male bird is not seen much in the vicinity of the nest until the young are hatched.

RANGE OF EGG COLOURATION AND MEASUREMENT: The eggs of the Bullfinch are usually four or five in number, sometimes as many as six. They are bluish-green or greenish-blue in ground colour, spotted with purplish-brown, and with paler underlying markings of brownish-pink. Some eggs are streaked with very dark brown, as well as spotted. The markings usually form a zone round the large end of the egg (sometimes round the small end), but in some specimens they are more uniformly distributed over the entire surface. Average measurement, ·75 inch in length, by ·55 inch in breadth. Incubation, performed by the female, lasts fourteen days.

DIAGNOSTIC CHARACTERS: The deep, clear blue of the ground colour, and their size, readily distinguish the eggs of the Bullfinch from those of any of the allied species breeding within the British area.

Family FRINGILLIDÆ.
Sub-family FRINGILLINÆ.

Genus PASSER.

HOUSE SPARROW.

PASSER DOMESTICUS (*Linnæus*).

Double Brooded.   Laying season, February to October.

BRITISH BREEDING AREA: Breeds more or less abundantly throughout the British Islands wherever houses are found, with one or two exceptions: these are in the wildest and most elevated districts of Scotland and Ireland. Most abundant in towns and well-cultivated districts.

BREEDING HABITS: The House Sparrow pairs for life, and during all the open months of the year may be found breeding in greater or lesser numbers. The great breeding season is in April, May, and June. There are few birds more gregarious than the House Sparrow, and the size of its breeding-colonies seems regulated purely by the extent of accommodation offered. No other British bird selects such a great variety of sites for its nest as the House Sparrow. It may be found almost everywhere—in every nook and crevice of all kinds of buildings, and amongst statuary, in trees (both in holes in the timber and in the open branches) and bushes, amongst ivy and other creeping plants, in holes in cliffs and sand-banks, both inland and marine, amongst the sticks of Rooks' nests, and even in deserted nests of Crows and Magpies. The materials used are just as varied in character, and it is difficult to name any soft substance that is not used at some time or another in the construction of the nest. Straws, dry grass, and herbage of all kinds, strips of rag, cotton, twine, worsted, wool, hair, and feathers are universally employed. The nests of the House Sparrow may be divided into two very distinct types, which differ considerably in form and in the quality of the workmanship. The first and commonest type, made in holes, is little more than a rude heap of material massed together, with the softest portions for the lining. The second type, placed in trees and amongst ivy, is much more skilfully made. Dry grass, straws, and withered plants are woven together into a large dome-shaped structure, with a small entrance hole on the top or side, and warmly lined with hair, wool, feathers, etc. These nests will be used year after year, and if destroyed, new ones will be made on the same sites, for the House Sparrow is greatly attached to its breeding-place. Both male and female assist in building the nest, and the thieving propensities of building birds often lead to combats. Brood after brood is reared in the same nest.

RANGE OF EGG COLOURATION AND MEASUREMENT: The eggs of the House Sparrow vary from four to six or even seven in number, five being the average clutch. They vary from bluish-white to pale grayish-brown in ground colour, more or less thickly mottled, blotched, and spotted with various shades of brown and gray. The eggs of this species present considerable variation, both as regards size, shape, and colour. Certain varieties are so thickly mottled and spotted that the ground colour is almost, if not entirely, concealed; others have the surface-spots small, ill-defined, and distributed over the entire surface of the shell; others have the spots fewer in number, but large and boldly defined; whilst others, yet again, have a zone of colour round the large or small end. In some the markings are chocolate-brown; in others, ash-brown or reddish-brown. Average measurement, ·9 inch in length, by ·6 inch in breadth. Incubation, performed by both sexes, lasts fourteen days. Very often the eggs are sat upon as soon as laid. It might also be remarked that in many instances one odd egg in a clutch is differently marked than the rest.

DIAGNOSTIC CHARACTERS: It is impossible to give any character by which the eggs of the House Sparrow may always be distinguished from those of allied species. Eggs of the Tree Sparrow are often indistinguishable; as are also eggs of the Pied Wagtail and the Meadow Pipit. As a general rule, the situation of the nest is a safe and unfailing guide to the identification of the eggs.

Family FRINGILLIDÆ. Genus PASSER.
Sub-family FRINGILLINÆ.

TREE SPARROW.

PASSER MONTANUS (*Linnæus*).

Double Brooded generally. Laying season, April to June or July.

BRITISH BREEDING AREA: Owing to wide-spread confusion with the House Sparrow, it is difficult to trace the breeding area of this species with any detailed completeness. It is certainly a local bird, and appears to breed most commonly in the eastern and midland counties of England, becoming rarer in the north and west. It breeds very locally in Wales, whilst in Scotland, although widely dispersed, it seems to be nowhere common. Its principal breeding area across the border is from the Lothians to Sutherlandshire on the east coast; although in the west, and on the Hebrides, it is by no means unknown, and I found it breeding even in remote St. Kilda. In Ireland it is even more local and sparingly distributed. It is difficult to say whether this species is slowly extending its range, or becoming better known and more universally distinguished from its commoner ally.

BREEDING HABITS: It is more than probable that the Tree Sparrow pairs for life, inasmuch as the same nesting-place will be used for a number of years in succession. In its choice of a breeding-haunt the Tree Sparrow differs considerably from the House Sparrow. Although it frequents the neighbourhood of farm-houses and out-buildings, it is much more of a field-haunting species, and in a great many cases rears its young in wild, uncultivated districts. The first nests are usually commenced towards the end of March. The site chosen varies considerably, and depends a good deal on the nature of the haunt. In some localities pollard willows are the favourite nesting-places; in others, holes in walls and cliffs, as I found to be the case in St. Kilda, the sides of old quarries, and even in the deserted nests of Crows and Magpies. I have also taken the nest from a hole in the branch of an oak tree. In other cases, a site is selected under the eaves of a building, or even in a hole in the thatch. In some few instances (when in old nests of Crows) the nest is domed and well made, but as a rule it is a slovenly structure, like that of all or most hole-builders, cup-shaped, and made of dry grass, straws, and roots, and warmly lined with feathers, and, less frequently, wool and hair. I have noticed that this species becomes very demonstrative when disturbed at the nest, and evinces much more anxiety than the House Sparrow usually does. Both male and female assist in the construction of the nest, in which as many as three broods are sometimes reared. The Tree Sparrow is nothing near so gregarious as the House Sparrow, and each pair of birds keep much to themselves.

RANGE OF EGG COLOURATION AND MEASUREMENT: The eggs of the Tree Sparrow are from four to six in number, and vary considerably in colour. Even in the same clutch one egg is often found much lighter in colour than the rest. They are grayish-white, or white with a faint blue tinge in ground colour, spotted and speckled with rich brown and grayish-brown, and with underlying markings of violet-gray. On some examples a few dark lines or streaks occur. The markings are generally so thickly distributed over the surface as to hide almost all trace of the pale ground colour; but on others, where the spots are larger and fewer, this is not the case, and then the gray underlying markings are also more conspicuous. Average measurement, ·79 inch in length, by ·55 inch in breadth. Incubation, performed by both sexes, but the greater part by the female, lasts fourteen days.

DIAGNOSTIC CHARACTERS: As a rule the eggs of the Tree Sparrow may be distinguished from those of the House Sparrow by their smaller size and more glossy texture. They do not vary so much in colour as the eggs of the latter species. In any case, however, they require careful identification, as the characteristics are too poorly defined to be absolutely reliable.

Family FRINGILLIDÆ.   Genus FRINGILLA.
Sub-family FRINGILLINÆ.

GREENFINCH.

FRINGILLA CHLORIS (*Linnæus*).

Double Brooded. Laying season, April to July, and even August.

BRITISH BREEDING AREA: Breeds more or less commonly in all the wooded districts throughout the British Islands. Its breeding area has been largely increased within the last fifty years, owing to the extensive planting of trees in many localities.

BREEDING HABITS: The principal breeding-haunts of the Greenfinch are in the well-cultivated districts, in shrubberies, parks, and gardens, and in the tall hedgerows of the lowland farms, in lanes, and on commons. The Greenfinch appears to pair annually, although I am of opinion that each pair of birds remain in company until two or three broods are reared, a new nest being made for each. The Greenfinch is remarkably social during the breeding season, and several nests may frequently be found quite close together. I have seen two nests in one small yew tree, and many similar instances have been recorded. The usual site for the nest of this species is in an evergreen of some kind, a dense thicket, or a tall hedge, whitethorn, perhaps, by preference. Other sites are frequently chosen however, as, for instance, in a gorse bush, amongst a cluster of woodbine, amongst ivy growing either on walls or trees, and less frequently fifty or sixty feet from the ground in a tall elm, either amongst the slender branches, or lower down, wedged closely into a crevice of the gnarled and knotted trunk. The nest varies a good deal according to locality. Some nests are made externally of twigs, others of coarse roots, intermixed with scraps of moss, and lined with finer roots, bits of wool, and quantities of hair and feathers. In other nests the external material is dry grass (I have known a nest made almost entirely of new-mown hay), twigs, and moss. The rim of the nest is generally well felted together, and the cup is smooth and neatly finished, although the lining varies a good deal in quantity and kind of material used. But little or no care is taken to conceal the nest, but generally the amount of surrounding foliage hides it from all but very close scrutiny. The old birds are quiet and seclusive during the nesting period, and the female is a close sitter, leaving the eggs with the greatest reluctance. Both male and female assist in making the nest. Some nests are much larger and better made than others.

RANGE OF EGG COLOURATION AND MEASUREMENT: The eggs of the Greenfinch are from four to six in number, and range from pure white to white tinged with blue or green in ground colour, somewhat sparingly speckled and spotted with reddish-brown, and with underlying markings of grayer brown. As a rule the spots are largest, most numerous, and deepest in colour at the large end of the egg, where they often form an irregular zone. In some eggs the pale underlying markings are large and more numerous than the surface-spots, and in others a few streaks or small spots of very dark brown occur. A rare variety is white and spotless. Average measurement, ·84 inch in length, by ·55 inch in breadth. Incubation, chiefly performed by the female, lasts fourteen days. Sometimes the first egg is sat upon as soon as laid.

DIAGNOSTIC CHARACTERS: It is impossible to give any character by which the eggs of the Greenfinch can be distinguished, either from those of the Crossbill, the Goldfinch, or the Linnet. As a rule the eggs are larger and more boldly spotted than those of the two latter species, and slightly smaller than those of the former. The style and situation of the nest are of some service in their identification.

Family FRINGILLIDÆ.   Genus FRINGILLA.
Sub-family *FRINGILLINÆ*.

GOLDFINCH.

FRINGILLA CARDUELIS, *Linnæus*.

Double Brooded occasionally. Laying season, May to July.

BRITISH BREEDING AREA: The Goldfinch is another bird whose breeding area has been sadly curtailed by the bird-catcher and modern improvements. Thirty years ago, in my part of Devonshire (Torquay) this charming bird bred in almost every orchard; now scarcely a pair can be found. It may still, however, be said to breed more or less sparingly and locally in all the English counties, becoming most rare in the extreme north and west. In Scotland it becomes even more local, although it is known to breed as far north as Caithness, and in some of the southern districts appears to be increasing in numbers. It has been known to breed on one occasion in Skye. In Ireland it is widely dispersed, but everywhere uncommon and very local.

BREEDING HABITS: The principal breeding-grounds of the Goldfinch are orchards and gardens, and the hedges and shrubberies near them. This bird appears to pair annually, although it may from time to time be seen in pairs all through the winter. The nest is usually made in a fork of some lichen-covered fruit tree; less frequently in an evergreen, and sometimes suspended from a drooping branch of a large tree. Occasionally it is made in a hedge near the garden or orchard. Few of our British nests equal that of the Goldfinch in beauty. It is almost as neatly made as that of the Chaffinch, the materials being felted in much the same way, but is smaller, and the garniture of lichens is not so conspicuous. It is made of moss, vegetable down, fine roots, and dry grass-stems, cemented with spiders' webs and a few bits of lichen, and lined with feathers, down, and hair. The cup is exquisitely finished, about two inches in diameter and one inch in depth. The old birds are remarkably quiet and careful not to betray the vicinity of the nest, the female sitting closely until approached.

RANGE OF EGG COLOURATION AND MEASUREMENT: The eggs of the Goldfinch are four or five in number, white, with a greenish or grayish tinge in ground colour, spotted and streaked with purplish-brown, and with underlying spots of gray. The amount of spots varies considerably, as also does their intensity of colour, some being almost black in appearance. As a rule most of the spots are displayed on the large end of the egg, where they often form a zone. A few long streaks or lines occasionally occur; and sometimes the eggs are almost devoid of markings. Average measurement, ·66 inch in length, by ·5 inch in breadth. Incubation, performed chiefly by the female, lasts fourteen days.

DIAGNOSTIC CHARACTERS: It is impossible to distinguish with certainty the eggs of the Goldfinch from those of the Siskin, but the locality of the nest is invariably different. From eggs of the Linnet and Greenfinch they may be usually separated by their smaller size, and from the latter by their smaller and more clearly-defined markings.

Family FRINGILLIDÆ.   Genus FRINGILLA.
Sub-family FRINGILLINÆ.

SISKIN.

FRINGILLA SPINUS, *Linnæus*.

Double Brooded. Laying season, April to June.

BRITISH BREEDING AREA: The Siskin is decidedly a northern species, and can only be regarded as breeding exceptionally in England. Instances are recorded of its having bred in Dorset, Sussex, Kent, Surrey, Middlesex, Oxfordshire, Gloucestershire, Bedfordshire, Derbyshire, Denbigh, Yorkshire, Westmoreland, Cumberland, and Durham. North of the border it becomes a more common bird, and breeds in many localities suited to its requirements, especially from Perthshire northwards to Sutherlandshire and Caithness, although chiefly in the eastern districts; but it is said to breed in Argyllshire. In Ireland its chief breeding area is in the eastern counties, in Antrim, Down, Wicklow, and Waterford.

BREEDING HABITS: The principal breeding-haunts of the Siskin are fir woods and plantations. The bird is said to frequent birch trees during this season, but this must be exceptional. The Siskin appears to pair annually. The nest is generally placed at a good height from the ground (from twenty to forty feet), either in the fork of a horizontal branch, or in a crotch near the top of the tree. In such a situation the nest is extremely hard to find, being difficult to see, owing to its small size and the density of the surrounding or intervening foliage. Both birds assist in its construction. The outside of the nest is made of fine twigs, grass-stalks, and roots, finally lined with moss, finer roots, vegetable down, hair, and occasionally feathers.

RANGE OF EGG COLOURATION AND MEASUREMENT: The eggs of the Siskin are five or six in number. They are very pale bluish-green in ground colour, spotted and speckled with dark reddish-brown, occasionally streaked with darker brown, and with underlying markings of violet-gray. As a rule most of the spots are on the larger half of the egg, and displayed more or less in a zone. Average measurement, ·65 inch in length, by ·51 inch in breadth. Incubation, performed almost entirely by the female, lasts fourteen days.

DIAGNOSTIC CHARACTERS: Unfortunately there is no character whatever to distinguish the eggs of the Siskin from those of certain other species. In the British Islands the eggs of the Goldfinch are most likely to be confused with them, but the structure and locality of the nest are safe guides to their correct identification.

Family FRINGILLIDÆ. Genus FRINGILLA.
Sub-family FRINGILLINÆ.

CHAFFINCH.

FRINGILLA CŒLEBS, *Linnæus*.

Double Brooded. Laying season, April to June, and even July.

BRITISH BREEDING AREA: Few of our British birds are more widely dispersed than the Chaffinch. It may be found breeding in all districts suited to its requirements throughout the British Islands, only being absent from the barest and treeless districts.

BREEDING HABITS: The principal haunts of the Chaffinch during the nesting season are gardens, orchards, spinneys, plantations, and hedgerows. This species pairs annually, and is one of the very first birds to do so, the love-song of the male commencing in February, or early in March. The nest, however, is rarely commenced before April. Almost every tree or good-sized bush in the summer haunts of the Chaffinch is destined at one time or another to contain its nest. The most general situations are in forks and crotches of the fruit trees, or in the tall bushes of the hedgerows; in whitethorn trees, or in the moss and lichen-covered branches of birches, elms, oaks, and ash trees. Less frequently the nest is placed in some evergreen, particularly a holly, or in a gorse bush. I have seen it made on a tuft of grass growing from a wall, and various other situations equally strange have been recorded. Perhaps with the sole exception of that of the Long-tailed Titmouse, the nest of no other British bird equals that of the Chaffinch in neatness and in beauty. A great variety of materials is used, and a series of the nests of this species not only present considerable variety in their appearance, but great diversity of skill, the best-finished nests probably being the work of old, experienced birds. Not only so, but the Chaffinch is careful to assimilate her nest to the colour of surrounding objects, so that materials suitable for one situation would be totally out of place in another. Moss, dry grass, fine roots, cobwebs, lichens, and wool are the materials principally used externally; feathers, hair, vegetable down, and wool are the usual lining. The external structure is subject to the most variation. Some nests are made externally almost entirely of green moss; others of green moss studded with different-coloured scraps of lichen, paper, or even decayed wood, attached firmly with spiders' webs; others are nearly all scraps of lichen. It will also be remarked that the garniture of lichen is often more abundant on one side than the other, and that the nest is shaped to the fork or crotch that supports it. The materials are well felted together. The nest is cup-shaped, the sides thick and substantial, and the hollow containing the eggs wonderfully neat and round. The female is the builder, but the male brings much of the material. A well-finished nest of the Chaffinch takes nearly a fortnight to build, and the parent birds are most solicitous for its welfare, even though but a scrap or two of material has been laid. A new nest is made for each successive brood. The Chaffinch displays no gregarious nor social tendencies during the breeding season, but sometimes several nests may be found at no great distance from each other.

RANGE OF EGG COLOURATION AND MEASUREMENT: The eggs of the Chaffinch are from four to six in number, the latter, however, being exceptional. They are pale bluish-green in ground colour, spotted and speckled, and sometimes streaked, with rich purplish-brown, and very often clouded or suffused with pale reddish-brown.

The underlying markings are paler brown. Many of the dark round spots are in the centre of paler markings One rather scarce variety is clear greenish-blue in ground colour, spotted and streaked (usually in a zone) with dark reddish-brown, and with paler brown underlying markings; another is very pale bluish-green, and entirely spotless. On some eggs the colouring matter is very eccentric, and displayed in one or two washes or splashes; on others it is confined to a circular patch on the large end of the egg, most intense in the centre, and shading off into paler tints. Average measurement, ·75 inch in length, by ·6 inch in breadth. The female alone incubates the eggs, the period lasting from twelve to fourteen days.

DIAGNOSTIC CHARACTERS: The eggs of the Chaffinch cannot readily be confused with those of any other species breeding in the British Islands. Their most striking characteristic is the round rich brown spots, situated on larger and paler markings. The eggs of the Brambling most nearly resemble them, but this species does not breed within the British area.

Family FRINGILLIDÆ. Genus LINOTA.
Sub-family FRINGILLINÆ.

LINNET.

LINOTA CANNABINA (*Linnæus*).

Double Brooded generally. Laying season, April to June.

BRITISH BREEDING AREA: Breeds in all districts suited to its requirements throughout the mainland of the British Islands, and in many parts of the Hebrides, where, like the Twite, it nests in the ling and heath.

BREEDING HABITS: The haunts affected by the Linnet during the breeding season are rough, uncultivated lands, especially on the borders of moors, gorse coverts, and hedgerows. By preference the bird frequents ground where gorse, broom, brambles, and low bushes occur. The Linnet pairs annually, somewhat early in the season, and at least a month before nest-building is commenced. The nest is made in a variety of situations. A favourite site is amongst gorse or broom; frequently a sapling fir is selected, and more rarely a crotch in a whitethorn or juniper. In mountain districts the nest is often made among the heather, quite close to the ground; and instances are on record where it has been found on the ground. Externally the nest is made of a few fine twigs (these are sometimes dispensed with) or dead sprays of gorse, grass-stalks, moss, occasionally a few dry leaves and bits of wool; internally it is well and warmly lined with hair, wool, vegetable down, and feathers, sometimes one and sometimes another of these materials predominating. The cup is beautifully rounded and neatly finished. As a rule the nest of the Linnet is made but a few feet from the ground, but I have seen nests at a considerable height above it. Both birds assist in its construction. The Linnet is a remarkably close sitter, but glides off the eggs with little demonstration. The cock sings throughout the period of incubation, and often denotes the vicinity of the nest by his habit of perching on some topmost spray near by.

RANGE OF EGG COLOURATION AND MEASUREMENT: The eggs of the Linnet are from four to six in number, the latter being perhaps the most frequently found. They are pale bluish-green in ground colour, sometimes greenish-white, finely spotted and speckled, chiefly round the large end, in the form of a zone, with deep reddish-brown and paler purplish-brown; the underlying markings of violet-gray are few and indistinct. Occasionally a few streaks or lines of very dark brown occur. Some eggs are much more sparingly spotted than others. Average measurement, ·7 inch in length, by ·53 inch in breadth. Both parents assist in the task of incubation, which lasts fourteen days, but the female sits the most.

DIAGNOSTIC CHARACTERS: It is impossible to distinguish certain small eggs of the Greenfinch from those of the Linnet, also small eggs of the latter from those of the Goldfinch. The construction and position of the nest is the safest guide to their identification. Eggs (and nest too) of the Twite resemble those of the Linnet so closely that great care is required in determining them correctly.

Family FRINGILLIDÆ.   Genus LINOTA.
Sub-family *FRINGILLINÆ.*

TWITE.

LINOTA FLAVIROSTRIS (*Linnæus*).

Single Brooded generally. Laying season, May and June.

BRITISH BREEDING AREA: The distribution of the Twite during the breeding season is dependent to a very great extent on the presence of moorlands. Wherever these districts occur, from the Midlands to the Orkneys and the Shetlands, and westwards to the Hebrides, the Twite may be found nesting upon them. It also breeds on St. Kilda. As may naturally be inferred, this species becomes most abundant in Scotland, especially in the west. The same remarks apply to Ireland, the bird breeding freely on all elevated moorlands.

BREEDING HABITS: The great and almost exclusive breeding-haunts of the Twite are the vast expanses of tall rank heather and ling. It is as common on the coast as in more inland localities, and nests freely on the small islands, narrow promontories, and sloping cliffs along the western coast-line of Scotland. The Twite pairs rather early, the flocks disbanding in spring and retiring from the lowlands to the moors during March and April. The usual site for the nest is amongst the long heather, only a few inches from the ground, and very frequently on the ground itself. A favourite situation is on the edge of a bank or ledge, especially by the side of a rough road or sheep-track. Less frequently the nest is made under a piece of turf or amongst long rank grass (as for instance at St. Kilda); and Capt. Elwes states that it is made in a bush or amongst creepers growing over walls. Twigs, dry grass-stems, moss, and roots form the outside of the nest, which is warmly lined with finer roots, vegetable down, wool, and feathers, each or all being used according to circumstances. The cup is round, deep, and very neatly finished. As a rule the nests placed above the ground are larger than those built upon it. Both birds assist in its construction. The Twite is a close sitter, and glides off her eggs with little demonstration; but very often after being driven from the nest she is joined by her mate, and the pair flit restlessly about until all is quiet again.

RANGE OF EGG COLOURATION AND MEASUREMENT: The eggs of the Twite vary from four to six in number, five being the average clutch. They are pale bluish-green in ground colour, finely spotted and speckled, and frequently streaked, with reddish-brown. Generally the bulk of the markings are distributed in a zone round the larger end of the egg. Some examples are much more streaked than others; and occasionally they are quite spotless. Average measurement, ·7 inch in length, by ·5 inch in breadth. Both parents assist in the duty of incubation, which lasts fourteen days, the female, however, sitting the most.

DIAGNOSTIC CHARACTERS: Absolutely indistinguishable from those of the Linnet. Perhaps they are on an average a little more streaked, but the most careful identification is required. The yellow bill, long tail, and absence of red from the crown and breast will serve to identify this species.

Family FRINGILLIDÆ.   Genus LINOTA.
Sub-family FRINGILLINÆ.

LESSER REDPOLE.

LINOTA RUFESCENS (*Vieillot*).

Single Brooded generally. Laying season, May and June.

BRITISH BREEDING AREA: The Lesser Redpole is another somewhat local bird during the breeding season. In England it breeds in most suitable localities north of lat. 52½°; but south of that limit it becomes not only local, but in many places rare or entirely absent, especially so in the extreme west and south-west. It certainly breeds in the lovely grounds of the Rock Walk at Torquay, and elsewhere in South Devon. In Scotland it is chiefly confined to birch woods and plantations of young firs. In Ireland it breeds in some numbers, but its principal haunts appear to be in the northern districts.

BREEDING HABITS: The Lesser Redpole frequents a variety of places for nesting purposes. In some districts it confines itself to the hedges in the fields; in others to small plantations and birch coppices. Frequently it has its haunt by the side of a mountain or moorland stream, amongst the alders and birches, and masses of bramble, or selects some quiet thicket in the more open woods or on commons. The flocks of Lesser Redpoles disband in spring, and each pair betake themselves to a favourite spot to nest. This Redpole pairs annually, and although the same nest is never revisited I have remarked considerable attachment to certain spots. The nest is generally built at no great height from the ground in a crotch in the hedge or sapling fir tree; at other times, however, it may be seen at heights of from twenty to fifty feet amongst the branches of tall trees. I have noticed the special partiality of this bird for an elevated site in an elm tree. The nests vary a good deal in size and quality of workmanship; the most exquisite examples I have ever seen being made in low whitethorn hedges. The external materials consist of slender twigs, roots, moss, and dry grass, and the lining consists of feathers, vegetable down, hair, and, less frequently, wool. The nest is cup-shaped, deep and round, and the cup is very smooth and neatly finished. The female sits closely, and seldom evinces much anxiety for her eggs.

RANGE OF EGG COLOURATION AND MEASUREMENT: The eggs of the Lesser Redpole are from four to six in number, five being the average clutch. They are greenish-blue in ground colour, spotted and speckled with purplish-brown, and sometimes streaked with dark brown. The underlying markings are pale brown or gray. Most of the markings are on the larger end of the egg. Average measurement, ·63 inch in length, by ·48 inch in breadth. Incubation, performed chiefly by the female, lasts fourteen days.

DIAGNOSTIC CHARACTERS: The deeply-tinted ground colour and remarkably small size of the eggs of the Lesser Redpole readily distinguish them from those of any other species breeding in our islands.

Family FRINGILLIDÆ. Genus PLECTROPHENAX.
Sub-family EMBERIZINÆ.

SNOW BUNTING.

PLECTROPHENAX NIVALIS (*Linnæus*).

Single Brooded.   Laying season, June and early July.

BRITISH BREEDING AREA: The Snow Bunting is a very local bird with us during the breeding season, and its nest has only been taken on few occasions in our islands. This may be to a great extent because the high mountain summits in some of the wildest parts of Scotland are rarely visited by naturalists. On the British mainland nests of the Snow Bunting have within comparatively recent years been discovered in Sutherlandshire; the bird has also long been known to breed on the Shetlands, notably in Unst and Yell.

BREEDING HABITS: In our islands the summer haunts of the Snow Bunting are amongst the wild rugged scenery of the north, either on the bare and rocky mountain sides and summits, or along the rough beaches of the sea, where drift-wood, loose stones, and rock fragments furnish shelter for the nest. It is not known whether the Snow Bunting pairs for life or forms a union each recurring spring, but probably the latter is the case. The nest is placed under loose

PLATE II.

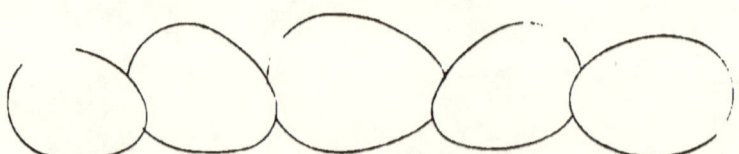

REED BUNTING. SNOW BUNTING. CORN BUNTING. CIRL BUNTING. YELLOW BUNTING

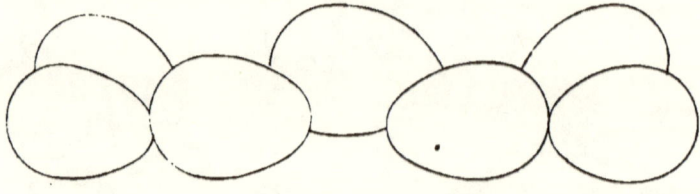

MEADOW PIPIT. WOOD LARK. SKYLARK. ROCK PIPIT. TREE PIPIT.

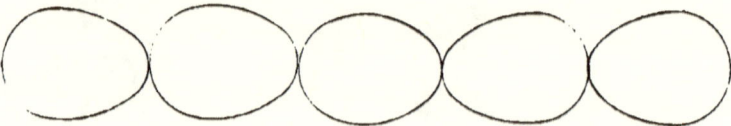

PIED WAGTAIL. WHITE WAGTAIL. GRAY WAGTAIL. BLUEH-WAGTAIL. YELLOW WAGTAIL

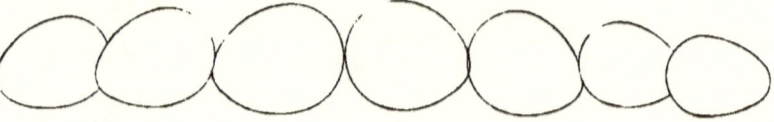

COAL TIT. MARSH TIT. GREAT TIT. BEARDED TIT. CRESTED TIT. BLUE TIT. LONG T-TIT

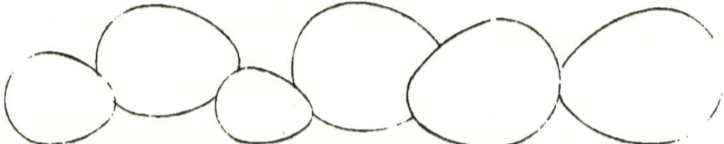

CREEPER. NUTHATCH. GOLDCREST. RED-BACKED SHRIKE. WOODCHAT SHRIKE.

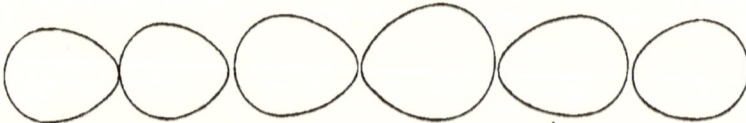

CHIFFCHAFF. WILLOW WREN. WOOD WREN. WHITE THROAT. DARTFORD WARB^R. L^R WIHTE THROAT.

stones in crevices of rocks, or amongst drift on the shore, and is generally well and carefully concealed. It is a bulky structure, but remarkably well and neatly finished inside. The outer materials consist of dry coarse grass, roots and moss, and the inside is lined with finer roots, hair (when obtainable), wool, and feathers. The female is a somewhat close sitter, but the male often reveals the presence of the nest by his behaviour in keeping close to the neighbourhood of the place.

RANGE OF EGG COLOURATION AND MEASUREMENT: The eggs of the Snow Bunting are from four to seven in number, but sometimes as many as eight have been found; five or six are an average clutch. They vary in ground colour from grayish-white to very pale blue, spotted and blotched most numerously on the larger end with rich reddish-brown, and occasionally pencilled with a few streaks and scratches of darker brown. The underlying markings are conspicuous, large, and numerous, and vary in tint from pale brown to pale gray. Zoned varieties are not uncommon. Average measurement, ·86 inch in length, by ·63 inch in breadth. Incubation, performed principally by the female, lasts about fourteen days.

DIAGNOSTIC CHARACTERS: The eggs of the Snow Bunting cannot readily be confused with those of any other species breeding in the British Islands, with the exception perhaps of those of the Corn Bunting; the smaller size and paler and smaller character of the markings will serve to distinguish them. It may be remarked that the eggs of the present species do not present the streaks and intricate pencillings that are so characteristic of the eggs of the Yellow Bunting and those of other allied species.

Family FRINGILLIDÆ.   Genus EMBERIZA.
Sub-family EMBERIZINÆ.

REED BUNTING.

EMBERIZA SCHOENICLUS, *Linnæus*.

Double Brooded.   Laying season, April to June.

BRITISH BREEDING AREA : The Reed Bunting breeds more or less commonly, if somewhat locally, in all suitable districts throughout the British Islands, extending even to the Hebrides and the Orkneys.

BREEDING HABITS : The principal haunts of the Reed Bunting are close to water of some description—disused brick- and clay-fields, the banks of slow-running rivers and streams, canals, and ponds ; and in wilder country the swamps and marshes in uncultivated districts. The Reed Bunting pairs early in spring, the male regaining his love-song towards the end of March or early in April. It is for the most part a solitary bird during the breeding season, living in pairs, and each pair keeping to a certain beat. The first nests are commenced about the middle of April in the southern districts, but nearly a month later further north. The nest is generally built on or quite close to the ground, but never suspended, as was once popularly supposed. It is usually built in the centre of a tuft of rushes, or beneath the shelter of the same, or well concealed amongst grass and other plants on the banks of the water. Hewitson has found the nest of this species, although rarely, on a bunch of reeds several feet above the water ; whilst Jardine remarked it not uncommonly in plantations of spruce fir, from three to nine feet from the ground. The nest is made externally of dry grass, moss, bits of withered reeds and flags, and internally of finer grass, hair, and the dry flowers of reeds. It is rather loosely put together, but well and neatly finished, and is made by both birds. The female sits very closely, and sometimes feigns lameness when rudely scared from the nest; a trick which the male occasionally joins his mate in performing. The latter frequently betrays the vicinity of the nest by his habit of perching conspicuously close by and warbling his monotonous song.

RANGE OF EGG COLOURATION AND MEASUREMENT: The eggs of the Reed Bunting are from four to six in number. They vary in ground colour from pale olive to buff, with a purple shade, boldly streaked and spotted with deep rich purplish-brown, some of them almost black in intensity. The underlying markings (both streaks and spots) are violet-gray. The spots are mostly paler and irregular in shape round the margin. The streaks are shorter, thicker, and not so net-like as those on the eggs of the Yellow Bunting. The markings are generally pretty evenly distributed over the entire surface of the shell, which is somewhat glossy. Average measurement, ·77 inch in length, by ·58 inch in breadth. Incubation, performed by the female, lasts fourteen days.

DIAGNOSTIC CHARACTERS: The comparatively short dark broad lines and large round spots are characters, in combination with their size, that readily serve to distinguish the eggs of the Reed Bunting from those of any other Bunting breeding in the British Islands.

Family FRINGILLIDÆ.   Genus EMBERIZA.
Sub-family EMBERIZINÆ.

CORN BUNTING.

EMBERIZA MILIARIA, *Linnæus*.

Single Brooded.   Laying season, May and June.

BRITISH BREEDING AREA: Although the Corn Bunting is widely distributed throughout the British Islands, it is decidedly a local bird. It is chiefly a lowland species, and is certainly most common on grain lands and rough pastures in maritime districts. It is found breeding most abundantly in the southern counties of England, becoming rarer and more local in the northern districts. In Scotland it is said to be most frequent in the southern, central, and western districts, reaching the Hebrides, but certainly not St. Kilda. One might just as much expect to find a Red Grouse in a flower-garden as a Corn Bunting in St. Kilda! It is tolerably well distributed throughout Ireland in suitable localities, but everywhere local.

BREEDING HABITS: Grain lands, sand dunes, and rough pastures are the favourite breeding-haunts of the Corn Bunting. It is not improbable that this species pairs for life, for I have known certain fields tenanted year by year for the purpose of nesting. If Corn Buntings are about they soon betray themselves by their monotonous notes, and habit of sitting for hours together on some tall weed or plant, occasionally flying a little way and then returning. The brown plumage of this species and its sluggish habits are very characteristic, and readily assist the observer to its identification. The nest is usually made upon the ground, but in some cases is situated amongst low brambles. Many nests are made on the ground amongst the mowing-grass or growing corn, or beneath a tuft of grass, or the shelter of a tall plant or bush. The nest, which is somewhat carelessly and loosely put together, is made externally of dry grass, moss, roots, "bull polls" or tufts of "twitch," and a few dead leaves, and lined with finer grass, roots, and hair. The female sits closely, until almost trodden upon in fact; but the male too often spoils his mate's precautions against discovery, by sitting poised on some stem, or even on the ground close by, droning out his monotonous song, and calling attention to its presence. The Corn Bunting is not at all social during the breeding season, and keeps in scattered pairs.

RANGE OF EGG COLOURATION AND MEASUREMENT: The eggs are from four to six in number. They vary in ground colour from pale buff to grayish-white, often tinted with purple; the surface-spots, streaks, and blotches vary from pale brown to very dark purplish-brown, and the underlying markings are various shades of violet-gray. Some eggs are much more boldly and intricately streaked than others; some have the blotches large and pale, distributed over most of the surface; some have the markings collected in an irregular zone round the larger end; whilst others are so thickly marked, either with surface-spots and streaks, or underlying markings, as to hide most of the ground colour. The eggs in each clutch are generally pretty uniform in character and appearance, however, in spite of the wide variation of colour and markings. Average measurement, ·98 inch in length, by ·7 inch in breadth. The female performs the greater part of the task of incubation, which lasts fourteen days.

DIAGNOSTIC CHARACTERS: The large size of the eggs of the Corn Bunting readily distinguish them from those of all other allied species breeding in the British Islands.

Family FRINGILLIDÆ.   Genus EMBERIZA.
Sub-family EMBERIZINÆ.

CIRL BUNTING.

EMBERIZA CIRLUS, *Linnæus.*

Double Brooded.   Laying season, May to July.

BRITISH BREEDING AREA: Few of our British birds have a more restricted breeding area than the Cirl Bunting. It breeds more or less sparingly and locally along the southern counties of England, from Cornwall to Kent; thence its range has been traced northwards through Surrey, Middlesex, Hertford, Bedford, and along the counties watered by the Thames to Gloucester, and in the counties of the Avon and Severn valleys to Warwick, Worcester, and Hereford.

BREEDING HABITS: The principal breeding-haunts of the Cirl Bunting are farm lands, gorse and bramble-covered commons, and the lanes and highways near them, in which plenty of trees and tall hedges occur. In Devonshire it frequents very similar haunts to those selected by the Yellow Bunting, although it is everywhere more partial to trees. The Cirl Bunting pairs towards the end of March or early in April. The nest, cunningly concealed, is usually placed in a low gorse bush, or amongst a thicket of briar and bramble. At other times it is built on a hedge-bank amongst the luxuriant growth of weeds, or wedged amongst the roots of the hedge-bushes. Generally, the nest is close to the ground, if not absolutely upon it, but occasionally it is situated as much as six feet above it. Externally it is made of roots, dry grass, and leaves, moss, "bull polls" or "twitch," and internally of finer roots and hair. According to circumstances the hair is omitted. It is somewhat loosely put together, but the interior is smooth and neatly finished. The female sits remarkably close, and quickly returns to the nest after being flushed.

RANGE OF EGG COLOURATION AND MEASUREMENT: The eggs of the Cirl Bunting are four or five in number; second clutches are not unfrequently composed of three. They are bluish-white in ground colour, sometimes with a faint greenish tinge, spotted and streaked with very dark brown, almost black. The streaks are bold, blotchy, and, as is usual, most frequent over the large end of the egg. The underlying markings, few, and faint in colour, are violet-gray. Average measurement, ·87 inch in length, by ·65 inch in breadth. Incubation, performed chiefly by the female, lasts fourteen days.

DIAGNOSTIC CHARACTERS: The intense, nearly black, lines and spots, and the bluish-white ground colour, readily distinguish the eggs of the Cirl Bunting from those of the Yellow Bunting, the only ones with which they are likely to be confused.

Family FRINGILLIDÆ.　　　　　　　　Genus EMBERIZA.
Sub-family *EMBERIZINÆ*.

YELLOW BUNTING.

EMBERIZA CITRINELLA, *Linnæus*.

Double Brooded.　Laying season, April to August.

BRITISH BREEDING AREA: The Yellow Bunting breeds more or less commonly in all districts suited to its requirements throughout the British Islands, even extending to the Outer Hebrides and the Orkneys, but not to the Shetlands.

BREEDING HABITS: The favourite haunts of the

Yellow Bunting are agricultural districts; wherever land is tilled, even in the wildest localities, the bird may almost invariably be met with during the nesting season. It also frequents in some numbers commons and uncultivated waste grounds, both on the uplands and the lowlands, and may be met with on the open spaces in coppices and well-timbered country. The Yellow Bunting pairs early in the year; the love-song of the male sounding persistently from the hedges and trees from March onwards to the following autumn. The nest is made in a variety of situations, both on the ground and at various heights above it. A favourite situation is on the bank of a hedge, either amongst the roots of the shrubs or bushes, or amongst the grass and other herbage. Another favourite spot is amongst nettles and other similar rank vegetation. Less frequently it is made in a gorse bush or a thicket of brambles and briars, and even more rarely in a sapling spruce fir. I have known the same spot used for several years in succession. Indeed the Yellow Bunting is much attached to a particular site, and frequently continues to lay egg after egg, even should the nest be removed. The nest is a rather bulky structure, loosely put together, but remarkably neat and well finished inside. Externally it is made of dry grass, bits of moss, roots and stalks of plants, and internally of finer roots and horsehair. Some nests, externally, are made almost exclusively of one or the other of the above-mentioned materials. A nest from Devonshire, now before me, is made entirely of straw and dry grass (obtained from the manure in a field near which it was taken), and lined with very fine roots and a few hairs—the whole stained very red from contact with the Devonshire soil. The hen sits closely, and leaves the nest in a silent manner, returning almost directly the nest is left in peace. Sometimes she tries various alluring antics to entice an intruder away.

RANGE OF EGG COLOURATION AND MEASUREMENT: The eggs of the Yellow Bunting are four or five in number, and vary in ground colour from white tinged with purple to purplish-brown. They are streaked, spotted, and flecked with dark purplish-brown, and with a few similar underlying markings of violet-gray. The amount and distribution of the pencilled markings vary considerably. On some eggs the marks are broad and bold, and occasionally expand into blotches; on others they are finer and interlaced in endless confusion, either over the entire surface or round the larger end, and emphasized here and there with broader and bolder scrawls. It is impossible to describe their endless pencillings, but it may be remarked that a strong family likeness runs through all eggs of this species. Average measurement, ·87 inch in length, by ·65 inch in breadth. Incubation, performed by both sexes, lasts fourteen days.

DIAGNOSTIC CHARACTERS: The distinct purple shade which suffuses the eggs of the Yellow Bunting, combined with the purplish-brown markings, and the elaborate and tortuous pencilling, are sufficient to distinguish them from those of all the allied species breeding in the British Islands.

Family ALAUDIDÆ.   Genus ALAUDA.

SKY-LARK.

ALAUDA ARVENSIS, *Linnæus.*

Double Brooded.   Laying season, April to July.

BRITISH BREEDING AREA: The Sky-Lark is generally distributed throughout the British Islands, breeding in every part, even including the Hebrides, the Orkneys and Shetlands.

BREEDING HABITS: The great breeding-grounds of the Sky-Lark are the farm lands; wherever agriculture is pursued the bird is present, and breeds in more or less abundance. It also frequents the upland moors and wild, uncultivated wastes, but is never found in woodlands or on country where trees are close together. Early in March the flocks of Sky-Larks begin to disband and to separate into pairs for the coming breeding season. During this period the Sky-Lark is not at all gregarious or even social, and each pair keeps exclusively to itself. Numbers of nests may often be found, however, within the area of fifty acres. The Sky-Lark appears to pair each spring; but not, I am of opinion, for every brood. During the mating season the male not only warbles on the ground as he runs to and fro, but may oft be seen in chase of the female, or beating off a rival for her favours. The nest is invariably built upon the ground, either amongst growing crops of grass and grain, amongst the coarse rank herbage on the common or waste, or snugly hidden in the ling and heath. It is made in a shallow depression, either scraped out by the bird or in the footprint of a horse or cow. It is not a very bulky structure, and is made of dry grass, straws, and scraps of moss, and lined with finer grass, fine roots, and a little horsehair. The female is a close sitter, and when leaving or visiting her nest usually drops into the cover some distance from it, and runs the remainder of the way through the grass or herbage. The best way to find nests of this species is to search systematically at dusk, walking up and down the fields until the birds are flushed from their eggs. They sit closer than usual at this time, and seldom rise until nearly trodden upon.

RANGE OF EGG COLOURATION AND MEASUREMENT: The eggs of the Sky-Lark are from three to five in number. They vary in ground colour from grayish-white to white tinged with olive, thickly spotted, mottled, and freckled with olive-brown, and with underlying markings of pale violet-gray. As a rule the markings are so thickly distributed as to conceal most of the ground colour; and round the larger end of the egg they very frequently run into a more or less clearly-defined zone, most distinct on eggs where the markings are not very profuse elsewhere. Rare varieties are white in ground colour, spotted and freckled with reddish-brown surface-spots, and lilac-gray underlying ones. The eggs of this species differ a good deal in size and shape, some being very rotund, others elongated. Average measurement, ·92 inch in length, by ·68 inch in breadth. Incubation, performed chiefly by the female, lasts fourteen days.

DIAGNOSTIC CHARACTERS: The numerous olive-brown markings which conceal most of the surface are a distinguishing characteristic of the eggs of the Sky-Lark; those of the Wood-Lark being spotted with reddish-brown. They most closely resemble those of the Crested Lark, but this species does not breed within the limits of the British Islands.

Family ALAUDIDÆ.   Genus ALAUDA.

WOOD-LARK.

ALAUDA ARBOREA, *Linnæus*.

Double Brooded, generally. Laying season, March to June.

BRITISH BREEDING AREA: The Wood-Lark is another very local, if somewhat widely distributed, species. Its principal breeding area in our islands is in the southern counties, from Devon to Kent, and northwards to Gloucestershire, Buckinghamshire, and Norfolk. North of these limits it becomes more local and rare, but probably breeds sparingly here and there, in suitable localities, as far as the Lake District. It is only known to have bred once in Scotland, in Stirlingshire, but probably has been overlooked. Its distribution in Ireland is very imperfectly known.

BREEDING HABITS: The favourite breeding-grounds of the Wood-Lark are well-timbered districts, where the soil is light, dry, and sandy. It is specially partial to heaths and commons, to fields on the borders of woods, and to parks; but everywhere the presence of trees seems essential to the Wood-Lark's requirements. The parties of Wood-Larks that have lived together during autumn and winter begin to disband early in spring, and to separate into pairs for the purpose of reproduction. The Wood-Lark appears to pair for life, and each season returns to some chosen spot in which to nest. In March the males may be seen toying with and chasing the females; and as soon as they have got back to their old haunts, the cock-birds repair to their accustomed perching-places on the tree-tops, close to where the nests are about to be made. The nest is almost invariably made upon the ground (although Professor Newton has recorded one in a stump of heather), sometimes in a very exposed situation on the bare turf, but more generally well concealed under brambles and briars, or beneath the shelter of a tuft of herbage or little bush. It is a simple structure, placed in a little hollow, usually scratched out by the parent birds, and made of coarse grass and moss, and neatly lined with finer grass and a little horsehair. Like so many other ground-building birds the Wood-Lark is a close sitter, and when leaving the nest voluntarily runs for some distance through the herbage ere taking wing. The male sings constantly and sweetly throughout the nesting season.

RANGE OF EGG COLOURATION AND MEASUREMENT: The eggs are four or five in number, and vary from pale buffish-white to white tinged with green in ground colour, spotted and freckled with reddish-brown, and with underlying markings of violet-gray. On some eggs the markings are pretty evenly distributed over the entire surface; on others they form a zone either round the end, or, more rarely, round the centre; whilst others have most of the colouring matter in a more or less confluent mass at either end. Average measurement, ·83 inch in length, by ·63 inch in breadth. Incubation, performed almost if not entirely by the female, lasts fourteen days.

DIAGNOSTIC CHARACTERS: The white ground colour and distinct reddish-brown markings distinguish the eggs of the Wood-Lark from every species breeding in the British Islands at all likely to be confused with them.

Family MOTACILLIDÆ.   Genus MOTACILLA.

PIED WAGTAIL.

MOTACILLA ALBA YARRELLII, *Gould.*

Double Brooded. Laying season, March to June.

BRITISH BREEDING AREA: The Pied Wagtail is widely and very generally distributed over the British Islands during the breeding season, extending in small numbers to some of the Hebrides and to the Orkneys, but not to St. Kilda, where I learned that it only occurred on passage.

BREEDING HABITS: The principal breeding-haunts of the Pied Wagtail are cultivated districts, those where water of some description is present being preferred. Brick-fields and the vicinity of clay-pits, where pools of water are numerous, are favourite localities, whilst the neighbourhood of country cottages and farm-houses, often far from open water, is equally preferred. It is very probable that some individuals of this species may pair for life. I stated the contrary in *Rural Bird Life* twelve years ago; but since then I have on various occasions known pairs of these Wagtails breed year after year in one particular hole of a wall or building. The nest is made in a great variety of situations, but generally in places well protected from view. Very frequently it is made in a hole in the wall of a cottage or outbuilding, or in a wall by the roadside, or by the side of a stream. Almost as frequently it may be found under a clod of earth or clay, or beneath a tile in the brick-fields; sometimes in a crevice of the stacks of unbaked bricks. Near clay-pits it is often placed far under a heap of clay blocks; whilst crevices in rocks or roots, or under steep bare river-banks, are utilized. A site is not unfrequently chosen in a hole of the thatch, or in the side of a wood- or hay-stack, or under a heap of stones in a quarry. The nest is a somewhat slovenly structure, but as a rule is well wedged and built into the hole. Externally it is made of almost any kind of vegetable refuse that may chance to be readily obtainable—dry grass, straws, twigs, roots, moss, dry leaves, bits of frayed rope or twine, wood-shavings, and large feathers, all loosely interwoven. The inside is thickly and warmly lined with any kind of hair that can be got, wool, and feathers. As a rule the frontage of the nest is much wider and more bulky than the back. The female sits rather closely. I have known her remain brooding on her eggs until a large heap of stones had been removed.

RANGE OF EGG COLOURATION AND MEASUREMENT: The eggs of the Pied Wagtail are from four to six in number, usually the latter. They are grayish or bluish-white in ground colour, thickly freckled, and more sparingly spotted with pale brown, and with numerous underlying markings of a similar character, of grayish-brown and violet-gray. On some eggs a few fine lines of very dark brown occur. There are at least two very distinct types. The first, usually grayish-white in ground colour, has the markings large and blotchy and pale brown, most of them on the surface, and round the large end of the egg. The second type has the ground colour bluish-white, and the markings finer, of a grayer brown, and dusted over the entire surface, but most numerous round the large end of the egg: the underlying markings are both small and numerous, and very gray in colour. Average measurement, ·8 inch in length, by ·6 inch in breadth. Incubation, performed chiefly by the female, lasts fourteen days.

DIAGNOSTIC CHARACTERS: It is impossible to give any character that will distinguish the eggs of the Pied

Wagtail from those of the White Wagtail: the latter species, however, breeds very locally and sparingly in our islands. Nor is it possible always to distinguish them from those of the House Sparrow. The situation of the nest (failing a sight of the parents) is the best guide to their identification.

Family MOTACILLIDÆ.   Genus MOTACILLA.

WHITE WAGTAIL.

MOTACILLA ALBA, *Linnæus.*

Double Brooded. Laying season, April to June.

BRITISH BREEDING AREA: Although the White Wagtail breeds commonly enough in France, Holland, and Belgium, on our side of the English Channel it is a rare and local bird, its place being taken by the well-known and widely-distributed Pied Wagtail. Instances of its breeding in our islands are, however, on record, and doubtless it may nest more frequently than is suspected. Has been met with breeding in Middlesex, Devonshire, the Isle of Wight, Kent, Huntingdonshire, and Cambridgeshire; and interbreeding with the Pied Wagtail in Norfolk and Suffolk.

BREEDING HABITS: So far as is known the White Wagtail does not differ in its habits, or in the localities it frequents, from the Pied Wagtail. Its nest is made of similar materials to those used by the commoner species, and is placed in much the same situations.

RANGE OF EGG COLOURATION AND MEASUREMENT: The eggs of the White Wagtail are from five to six

or even seven in number, and vary much more than
those of the Pied Wagtail. I attribute this peculiarity
to the fact of the Pied Wagtail having a very restricted
geographical area, and being decidedly the youngest
species. They vary in ground colour from pure white
to very pale blue, spotted, freckled, and splashed with
grayish-brown and various shades of reddish-brown.
The underlying markings of the same general character
are various shades of gray. Occasionally a few hair-like,
nearly black streaks occur, notably on the large end.
The spots are sometimes small and distributed over the
entire surface; at others they are chiefly collected in a
broad zone round the large end of the egg; whilst
others, yet again, are distributed in splashes and
mottlings over the entire surface. Average measurement, the same as in the preceding species. Incubation
period the same.

DIAGNOSTIC CHARACTERS: It is impossible to give
any character by which the eggs of the White Wagtail
may be distinguished from those of the Pied Wagtail;
the range of variation is wider, and the markings are
richer and larger. The White Wagtail is easily identified by its slate-gray instead of black upper plumage
below the nape.

Family MOTACILLIDÆ. Genus MOTACILLA.

GRAY WAGTAIL.

MOTACILLA SULPHUREA, *Bechstein*.

Probably Double Brooded. Laying season, April to June.

BRITISH BREEDING AREA: The Gray Wagtail is pretty generally distributed in the mountain districts throughout the British Islands, becoming more local in the south of England. It however breeds more or less sparingly on Dartmoor, Exmoor, along the district of the Downs, and Salisbury Plain. From the Peak of Derbyshire northwards through Scotland (including some of the Inner Hebrides) the bird becomes commoner and more generally dispersed, because districts suited to its requirements are more frequent. The breeding area of this Wagtail in Ireland is very imperfectly known.

BREEDING HABITS: The favourite, almost the only, breeding-grounds of the Gray Wagtail in our islands are the rough wild banks of the rapid mountain streams, which are fringed with alders, birches, and the mountain ash, and clothed almost to the water's edge with a great variety of more lowly vegetation,—long coarse grass, ferns, bracken, briars, brambles, and such like. As the Gray Wagtail may be found nesting in one particular spot season after season, we may reasonably conclude that this species pairs for life. The Gray Wagtail begins to draw near its upland haunts from the lowlands where it spends the winter, in February and March. Each pair appears always to journey in company, and to betake themselves to the particular part of the mountain stream in which they have a vested right. They are not social birds, and keep very closely to certain reaches of the stream. The site for the nest is never far from the water-side; in many cases but a few inches from the stream. Sometimes the nest is made under an overhanging rock on the bank, amongst tall grass and weeds; sometimes a large stone, almost covered with bramble and fern, conceals it; at others it is placed amongst drifted rubbish, or on the low stump of a tree. I have known it to be made in a hole in a dry wall; in fact almost any well-sheltered nook by the stream is utilized. A nest of this species has been recorded as being built on an old deserted nest of the Song Thrush. The nest of the Gray Wagtail is loosely put together, and made externally of roots, dry grass, and moss, and lined with finer roots, cow- or horsehair, and, less frequently, feathers. The female sits closely, and when disturbed often flies into the nearest tree, when she is usually joined by her mate, and the two birds are very restless until left in peace.

RANGE OF EGG COLOURATION AND MEASUREMENT: The eggs of this species are generally five in number, sometimes four, more rarely six. They vary from grayish-white to yellowish or buffish-white in ground colour, mottled and speckled with pale brown of various shades, and occasionally marked with a few irregular lines of very dark brown. Average measurement, ·75 inch in length, by ·55 inch in breadth. Incubation, performed chiefly by the female, lasts thirteen or fourteen days.

DIAGNOSTIC CHARACTERS: It is impossible to give any reliable character by which the eggs of the Gray Wagtail may be distinguished from those of the Yellow Wagtail and the Blue-headed Wagtail, or even from those of the Sedge Warbler. The situation of the nest, and the nature of the country, however, prevent much chance of confusion.

Family MOTACILLIDÆ.   Genus MOTACILLA.

BLUE-HEADED WAGTAIL.

MOTACILLA FLAVA, *Linnæus*.

Probably Single Brooded.   Laying season, May and June.

BRITISH BREEDING AREA: The Blue-headed Wagtail, although common enough across the Channel, is even more rarely known to breed within our limits than the White Wagtail, although there can be small doubt that nests are repeatedly overlooked. The only authentic instances of this species breeding in the British Islands, so far as is known, were near Gateshead, where its nest has been taken several times.

BREEDING HABITS: The principal haunts of the Blue-headed Wagtail during the breeding season are marshy meadows and pastures. The flocks of this species that have kept company during the winter disband in April and separate into pairs. The Blue-headed Wagtail appears to pair annually. The site for the nest is invariably on the ground, usually on a bank of the hedgerows, or amongst the herbage of the meadows. The nest, loosely put together, is made externally of dry grass, roots, and bits of moss, and lined with finer roots, horsehair, and occasionally a few feathers. Like most ground-building birds, the present species is a close sitter.

RANGE OF EGG COLOURATION AND MEASUREMENT: The eggs of the Blue-headed Wagtail are usually five in number, but sometimes as many as six, or in rarer instances only four. They vary from yellowish-white to very pale bluish-white in ground colour, mottled, freckled, and clouded with pale brown. Occasionally a few very dark hair-like lines occur on the larger end of the egg. Average measurement, ·78 inch in length, by ·56 inch in breadth. Incubation, performed chiefly by the female, lasts about fourteen days.

DIAGNOSTIC CHARACTERS: As previously remarked, the eggs of the Blue-headed Wagtail so closely resemble those of the Gray Wagtail and the Yellow Wagtail, that no character can be given by which they can be separated. The eggs require careful identification; but it should be remembered that the present species is very rarely known to breed within our area.

Family MOTACILLIDÆ.   Genus MOTACILLA.

YELLOW WAGTAIL.

MOTACILLA RAII, *Bonaparte*.

Double Brooded. Laying season, April and June.

BRITISH BREEDING AREA: A common summer visitor, widely distributed during the breeding season throughout England, with the exception of Cornwall and Devon. In Scotland its principal breeding area lies south of Stirling. In Ireland it becomes even more local, and is only known to nest in the neighbourhood of Lough Neagh, and the vicinity of Dublin.

BREEDING HABITS: Marshy meadows and pastures, broads, and commons are the favourite breeding-grounds of the Yellow Wagtail. In this respect it closely resembles the Blue-headed Wagtail; in fact the habits of the two birds, like those of the Pied and White Wagtails, are almost precisely alike. This Wagtail appears to pair annually, shortly after its arrival in spring; although it is probable that some pairs remain united for life, as we frequently find a certain spot tenanted year by year. The nest, made towards the end of April, is invariably placed on the ground, and generally well concealed, either by a tuft of herbage, a projecting stone, or a clod of earth. It is made of a variety of materials, those being selected that are most readily obtainable. As a rule, dry coarse grass, roots, and moss form the outside; the inside is somewhat neatly lined with finer roots and feathers, hair and fur, according to circumstances. The nest is rather loosely put together.

RANGE OF EGG COLOURATION AND MEASUREMENT: The eggs of the Yellow Wagtail are from four to six in number. They are grayish-white in ground colour (exceptionally suffused with a rosy tinge), mottled and speckled with pale yellowish-brown or olive-brown, and occasionally streaked with blackish-brown on the larger end. The colouring matter is generally so thickly distributed over the entire surface of the egg as to hide all or nearly all the pale ground colour. Average measurement, ·78 inch in length, by ·56 inch in breadth. Incubation, performed chiefly by the female, lasts about fourteen days.

DIAGNOSTIC CHARACTERS: As previously remarked, the eggs of the Yellow Wagtail are often absolutely indistinguishable from those of the Blue-headed and Gray Wagtails, and from certain varieties of those of the Sedge Warbler. They require most careful identification.

Family MOTACILLIDÆ. Genus ANTHUS.

TREE PIPIT.

ANTHUS TRIVIALIS (*Linnæus*).

Single Brooded generally. Laying season, May and June.

BRITISH BREEDING AREA: With the exception of Wales and the extreme south-west (Cornwall and parts of Devon) the Tree Pipit is a widely and commonly distributed species in England, due allowance of course being made for the suitability of the district to its requirements. In Scotland it becomes much more local, its principal strongholds being in the south, especially in the Solway area and near Glasgow. It becomes much rarer in the north, although it has been found breeding as high as Sutherlandshire. The evidence for this species having bred in Ireland is unreliable.

BREEDING HABITS: The Tree Pipit is a summer visitor to our islands, arriving in April. Its principal haunts during the nesting period are well-timbered farm lands, the more open spaces in old forests, the borders of woods and plantations, and in parks. It is not improbable that this Pipit pairs for life, as season after season the old haunts are tenanted, and even certain trees are used year by year as perching-places for the males, during the intervals of their song-flights. The nest, invariably built upon the ground, is generally well concealed amongst herbage, although not unfrequently it may be found in very exposed situations, such as on the short turf under a tree in the pastures. Sometimes it is made on an open bank in the woods, or amongst growing grass, clover, grain, and such-like crops. Banks below hedges, either in the fields or by the wayside, are sometimes selected, as is also the shelter of a little bush or grass tuft on the common. The nest is made in a little hollow scraped out by the parent birds, and is formed externally of dry grass, twitch, moss, roots, and internally of finer grass, roots, and horsehair. It is not a very large structure, and is usually loosely put together, but neatly finished inside. Many nests of this species contain no horsehair, but are lined with fine grass alone. The Tree Pipit is a close sitter, and usually visits and leaves its nest by running for some distance through the herbage before taking wing. The best way to find the nests of this species is to walk up and down the ground at dusk. The approximate locality of a nest may be generally determined by the persistent presence of the cock-bird in some tree close by.

RANGE OF EGG COLOURATION AND MEASUREMENT: The eggs of the Tree Pipit are from four to six in number, a clutch very frequently being of the latter amount. Few eggs of our British Passeres vary so much as those of the present species. They range in ground colour from grayish-white and bluish-white to pinkish-white and pale olive, mottled, spotted, and blotched with various shades of reddish-brown, purplish-brown, and olive-brown, and occasionally streaked with irregular lines of nearly black. The distribution and character of the markings vary considerably. Some eggs are so densely mottled and spotted as to conceal all or nearly all the ground colour; others have the colouring matter mostly in a round patch or in an irregular circle round the larger end; whilst others, yet again, are pretty evenly covered with splashes, blotches, and round spots, darkest in the centre. Eggs in the same clutch are invariably of the same colour and type. I have often remarked that the darkest-coloured eggs are generally found in the shady situations, whilst those of lighter tints are found in nests made in bare and open localities. Average measurement, ·82 inch in length, by ·6 inch in breadth. The female performs most of the duties of incubation, which lasts thirteen or fourteen days.

DIAGNOSTIC CHARACTERS: The pinkish or reddish-brown appearance of the densely-marked eggs, and the bold spots and blotches of the olive types, are generally sufficient to distinguish the eggs of the Tree Pipit from those of the Meadow Pipit; the eggs of the latter bird are also smaller and more constantly browner. The eggs of the Rock Pipit are sometimes met with of a reddish tint; but olive-browns prevail. The Rock Pipit also never breeds inland, only on rocky coasts.

Family MOTACILLIDÆ. Genus ANTHUS.

MEADOW PIPIT.

ANTHUS PRATENSIS (*Linnæus*).

Single Brooded generally. Laying season, April to June.

BRITISH BREEDING AREA: The Meadow Pipit is by far the commonest and most widely distributed species of this genus breeding in the British Islands. It breeds more or less commonly in almost every part of the United Kingdom, including the Channel Islands in the south, the Orkneys and Shetlands in the north, the Hebrides, St. Kilda, and the Blasket Islands in the west.

BREEDING HABITS: Few British birds breed more generally throughout our islands, on uplands and lowlands alike, than the unobtrusive Meadow Pipit. It may be found nesting almost everywhere—on moorlands and mountains, on sea-girt islets, on commons, pastures, fields, parks, and open grounds in coppices, by the water-side, on the country highways, and even in swamps. The Meadow Pipit appears to pair annually and very early in the season. The nest is invariably made upon the ground, under the shelter of a stone, a tuft of herbage, or a little bush, and very frequently amongst the wet moss on the marshes, or in the centre of a tuft of rushes. It may also be found very often amongst the long grass on a bank, or in heather and ling. The nest is simple and loosely put together, made in the first place of moss, dry grass, and bits of heath and reed, and lined with finer grass, fine roots, and horsehair. The Meadow Pipit sits closely, usually not leaving her nest until it is nearly under our feet. This species cannot be described as social during the breeding season (although it is gregarious enough during autumn), but numbers of nests may be met with in a comparatively small area.

RANGE OF EGG COLOURATION AND MEASUREMENT: The eggs of the Meadow Pipit are four or five in number, occasionally six. They are white, suffused with brown or pale bluish-green, in ground colour, mottled, spotted, and speckled with various shades of brown, and with paler brown underlying markings. Sometimes a few hair-like streaks of blackish-brown occur chiefly on the larger end of the egg. Usually the markings are so closely distributed that little of the ground colour is visible; but in others the spots are larger, fewer in number, and sometimes form a zone round the larger end of the egg. Average measurement, ·79 inch in length, by ·58 inch in breadth. Incubation, performed mostly by the female, lasts fourteen days.

DIAGNOSTIC CHARACTERS: The eggs of the Meadow Pipit may be readily identified by their small size and brown appearance.

Family MOTACILLIDÆ.   Genus ANTHUS.

ROCK PIPIT.

ANTHUS OBSCURUS (*Latham*).

Double Brooded.   Laying season, April to July.

BRITISH BREEDING AREA: The Rock Pipit breeds more or less commonly on all the rocky coasts of the British Islands, including the Channel Islands in the south, the Orkneys and Shetlands in the north, the Hebrides, St. Kilda, and the Blaskets in the west, and the Farne Islands and Bass Rock in the east.

BREEDING HABITS: The breeding-grounds of the Rock Pipit are always situated near the sea. This Pipit is a thorough rock bird during the nesting season, and never breeds save on or near a rock-bound coast. Although this species cannot be classed as gregarious or even social during summer, many pairs of birds may be met with along a short distance of suitable coast; and the smallest of rocky islets frequently contains several pairs. The Rock Pipit pairs very early in spring; nevertheless it is by no means an early breeder. The nest, which is generally by no means easy to find, unless stumbled upon by accident, is invariably placed on the ground, and may be met with in a great variety of situations. The site is usually a sheltered one, more or less, either under a mass of rock or a big stone, under a heap of dry sea-weed, or in some crevice of the cliffs, it may be hundreds of feet sheer above the water. Other nests may frequently be found amongst dense beds of sea-campion, or sheltered by a tuft of sea-pinks. I have taken the nest of this species from a disused Puffin's burrow, and from a hole in the walls of a ruined hut. The nest is never made far from the water.

Externally it is made of dry grass, sea-weed, and dry scraps of marine vegetation; sometimes moss is almost exclusively used; internally it is lined with finer grass and horsehair, wherever such can be obtained. I once found a nest of this species at the Farne Islands containing one large white Gull's feather in the lining. At St. Kilda the birds pull the Puffin snares to pieces to obtain the horsehair. The whole structure is loosely put together, but the inside is neatly arranged. The Rock Pipit is a close sitter, but when scared from the nest both parents frequently flit up and down in a very restless, anxious manner.

RANGE OF EGG COLOURATION AND MEASUREMENT: The eggs of the Rock Pipit are four or five in number. They vary in ground colour from pale greenish-gray to pale brownish-gray, or nearly white, mottled, spotted, and blotched with olive-brown or reddish-brown, and with underlying markings of grayish-brown. On some eggs a few dark brown lines or streaks occur, usually at the larger end. The markings are generally pretty evenly distributed over the entire surface; but occasionally they are most numerous, and form a zone round the larger end of the egg, and sometimes they are fewer and larger. Average measurement, ·85 inch in length, by ·63 inch in breadth. Incubation, performed chiefly by the female, lasts thirteen or fourteen days.

DIAGNOSTIC CHARACTERS: The large size and brown appearance of the eggs of the Rock Pipit, combined with the situation of the nest, serve to distinguish them from those of other British species most likely to be confused with them.

Family CERTHIIDÆ. Genus CERTHIA.

COMMON CREEPER.

CERTHIA FAMILIARIS, *Linnæus*.

Double Brooded. Laying season, April to June.

BRITISH BREEDING AREA: The Creeper is more or less commonly distributed throughout the woodland districts of the British Islands, but is absent as a breeding species from the Hebrides (except Skye, where it is a local resident), the Orkneys and the Shetlands, and other treeless areas.

BREEDING HABITS: The Creeper is a resident bird in the British Islands. Its favourite haunts are districts full of large trees, especially localities where the timber is ancient and more or less decayed. For this reason the bird is most abundant in old forests, extensive woods, and well-timbered parks. Less frequently it may be met with during the breeding season in large orchards, especially such where the trees are old and abound in hollows and crevices. There can be little doubt that the Creeper pairs for life, and at all times may be met with in the company of its mate. The nest of this species is made in a variety of situations, perhaps most frequently in a crevice where a strip of bark has peeled or been torn away from the trunk of a tree. Occasionally a hole in the trunk is selected. In rare instances the nest is made in the side of a stack of cord-wood, in a hole of a building, or under thatch. Instances are on record where it has been met with in a deserted Hawk's nest, and even in a pile of bricks. The nest of the Creeper varies a good deal in bulk, according to the capacity of the selected site, most of the surplus space being filled with a mass of fine twigs, the rim of the nest being formed of the same material; the interior is made of fine roots, dry grass, strips of inner bark, and moss, and frequently finished off with wool and feathers in small quantities. Almost every kind of forest tree will be selected, provided a suitable site can be obtained. The Creeper is neither social nor gregarious during the breeding season, each pair keeping entirely to themselves. The female sits closely, and when disturbed slips quietly away without demonstration.

RANGE OF EGG COLOURATION AND MEASUREMENT: The eggs of the Creeper are from six to nine in number, white, or white tinged with yellow in ground colour, spotted and blotched with reddish-brown or brownish-red, and with underlying markings of similar character, but violet-gray in colour. The markings differ a good deal in size and distribution; some being minute and very dark in colour, others blotchy and paler; some confined to a zone round the larger end of the egg, others equally distributed over the entire surface. On some eggs the markings are very few and minute. Average measurement, ·62 inch in length, by ·47 inch in breadth. Incubation, performed chiefly by the female, lasts from fourteen to fifteen days.

DIAGNOSTIC CHARACTERS: The eggs of the Creeper are indistinguishable from those of some of the Tits and allied species, but the singular nest affords an easy and constantly reliable means of identification.

Family PARIDÆ.
Sub-family SITTINÆ.

Genus SITTA.

COMMON NUTHATCH.
SITTA CÆSIA, *Wolf.*

Probably Double Brooded. Laying season, April to June and July.

BRITISH BREEDING AREA: The Nuthatch is another very local species, confined principally to England, being very rare and accidental in Scotland, and entirely absent, so far as is at present known, from Ireland. It is most commonly dispersed in the southern and central counties, breeding certainly as far west as Devonshire. In the north of England it is much more local and rare, and apparently does not breed north of Yorkshire, whilst in Wales, where it is said to be on the increase, it certainly nests in many localities, in Breconshire and elsewhere.

BREEDING HABITS: The breeding-haunts of the Nuthatch are woods, especially where plenty of old timber is to be found, old orchards, and country where large timber is plentiful. It frequently nests in large game coverts and plantations; less frequently in barer situations, and occasionally near houses. The Nuthatch pairs for life. All the winter through it may be observed in pairs, and returns season after season to one particular spot to breed. It is not by any means a social bird, each pair keeping to one particular haunt. The usual site for the nest is in a hole in a tree trunk, or a large branch, or in a stump. More rarely a hole in a wall or a cavity in the side of a haystack is utilized. The distance from the ground seems purely to depend on the site, little if any choice being exercised. The entrance is invariably plastered up with mud or clay, and a neat round hole formed for ingress. The amount of clay varies a good deal according to circumstances: if the cavity is large the entrance is reduced to just sufficient for the passage of the birds. As many as eleven pounds of clay have been exceptionally found attached to a nest in the side of a haystack.[1] Beyond this more or less elaborate structure of clay the nest of the Nuthatch is crude and simple, and so loosely fabricated as to render its removal entire an impossibility. It is almost invariably composed of a layer of dry leaves and flakes of bark, arranged at the bottom of the hole. Dry grass is used, but only very exceptionally. It should also be remarked that the Nuthatch sometimes enlarges a hole, and then a good deal of wood-dust and chips are mixed with the usual materials. The depth of the holes varies considerably, from a few inches to a foot, or even more. The Nuthatch is not easily driven away from its nesting-hole, and will replace the clay at the entrance after it has been removed repeatedly, and return year after year in spite of regular disturbance. It will also continue laying egg after egg after they are taken. The Nuthatch sits very closely, usually allowing itself to be removed by the hand from the nest-hole rather than desert its eggs.

RANGE OF EGG COLOURATION AND MEASUREMENT: The eggs of the Nuthatch are from five to eight in number. They are pure white in ground colour, spotted, and more rarely blotched with reddish-brown, and with similar but fewer underlying markings of violet-gray The distribution of the spots varies considerably. On some eggs they are uniformly distributed over the entire surface; on others they form an irregular zone or a circular patch on the larger end; more exceptionally they are handsomely blotched, and even more rarely

[1] This nest is now in the Natural History Museum at South Kensington.

still a few short lines of very dark brown occur. Average measurement, ·78 inch in length, by ·57 inch in breadth. Incubation, performed pretty equally by both sexes, lasts fourteen days.

DIAGNOSTIC CHARACTERS: The eggs of the Nuthatch cannot with safety be distinguished from those of some allied species, but the character of the nest is an unfailing means of correct identification.

Family PARIDÆ.   Genus PANURUS.
Sub-family *PANURINÆ*.

BEARDED TITMOUSE.

PANURUS BIARMICUS (*Linnæus*).

Double Brooded. Laying season, April to June and July.

BRITISH BREEDING AREA: No more local bird breeds within the British Islands than the Bearded Titmouse. Formerly it was much more widely distributed than is now the case, the drainage of fens and the persecution of collectors having well-nigh exterminated it from all its accustomed haunts. Almost its only stronghold now is the district of the Broads in Norfolk. It may possibly still continue to breed sparingly in Suffolk, as it certainly does in at least one locality in Devonshire.

BREEDING HABITS: The favourite, and we may safely say the exclusive, haunts of the Bearded Titmouse are extensive reed-beds—the vast expanses of marsh and reed-clothed waters that may now almost only be found in one English county. I am unable to say whether the Bearded Titmouse pairs annually or remains in company with its mate for life; probably it is the latter. The nest is made amongst the vegetation of the marshes, but never suspended from stems of reeds, like that of the Reed Warbler. It is usually built under a tuft of sedge or aquatic herbage, sometimes as much as a foot or more from the ground, and invariably well concealed by the overhanging and surrounding vegetation. Externally it is made of dry grass, leaves, bits of reed, and other scraps of withered aquatic vegetation; internally, of finer grass and the flowers of the reeds. It is loosely put together, but neatly finished. The Bearded Titmouse is a close sitter, and when flushed from the eggs slips silently away and hides amongst the vegetation; but if the nest contain young the old birds become more demonstrative. The Bearded Titmouse is not gregarious during the breeding season, but as evidence of its social tendencies during this period, it may be remarked that two females have been said to occupy the same nest. This species is evidently very prolific, as two females (mated to one male), kept in confinement, laid forty-nine eggs in about a couple of months!

RANGE OF EGG COLOURATION AND MEASUREMENT: The eggs of the Bearded Titmouse vary from five to seven in number. They are creamy-white in ground colour, marked with short irregular lines, and more sparingly with specks of dark brown. They are subject to little variation. Average measurement, ·7 inch in length, by ·55 inch in breadth. Incubation, performed chiefly by the female, lasts fourteen days.

DIAGNOSTIC CHARACTERS: The size of and peculiar streaky or line-like markings on the eggs of the Bearded Titmouse distinguish them at a glance from those of every other species breeding in the British Islands.

Family PARIDÆ   Genus ACREDULA.
Sub-family PARINÆ.

LONG-TAILED TITMOUSE.

ACREDULA CAUDATA ROSEA, *Blyth*.

Single Brooded generally. Laying season, April and May.

BRITISH BREEDING AREA: The Long-tailed Titmouse is pretty generally, if somewhat locally, distributed throughout all districts suited to its requirements in England and Wales. In Scotland it becomes rarer and more local, owing to suitable localities not being so numerous, although it breeds pretty generally in the wooded districts, even extending to Islay and Skye, in which latter island I met with it some years ago. The breeding range of this bird in Ireland is very imperfectly defined, but extends, it is said, to most wooded areas.

BREEDING HABITS: The favourite breeding-grounds of this Titmouse are plantations, woods, shrubberies, commons covered with thicket and underwood, and tall thick hedges on farm lands. It is also partial to an orchard or a large garden, and, less frequently, extensive gorse coverts. I am of opinion that the Long-tailed Titmouse pairs for life; and although the same nest is not used annually, or in many cases not even the same locality, the bird may be seen in pairs all through the year. Although gregarious to some extent during the remainder of the year, there is no social tendency in the breeding season. The nesting-site varies considerably, both in situation and in altitude, sometimes being as low as five feet from the ground, at others as much as fifty feet above it. Almost every kind of bush or tree may be selected, but preference is perhaps shown for those of an evergreen nature. Firs, hollies, yews, and gorse are perhaps the special favourites, but brambles, ivy, woodbine, hazels, whitethorns, and birches are frequently selected, as are also the branches of oaks, elms, and other forest trees. The nest of this species is one of the handsomest examples of bird-architecture to be found in the world, and takes about a fortnight to complete. It is globular in shape, like that of the Wren, with a small hole in one side near the top for ingress. The outer materials are chiefly composed of moss and lichens, cemented and felted together with spiders' webs, and often bits of wool; the interior is thickly and warmly lined with feathers and hair, the former usually predominating. The substance of the nest is very similar to that of the Chaffinch, and the outside is generally made closely to resemble surrounding objects, with a view to concealment. I have on one occasion had a nest of this species with a kind of flap over the entrance hole, which must have been raised every time the parents either entered or quitted the structure. Selby and others have asserted that the nests contain two means of exit, but if this statement be true, the fact must be a very exceptional one.

RANGE OF EGG COLOURATION AND MEASUREMENT: The eggs of the Long-tailed Titmouse are usually from six to eight or ten in number, but instances are on record where many more have been found, up to sixteen and even twenty. It is probable that these very large clutches are the produce of two females. I have certainly seen several birds near one nest on more than one occasion. The eggs of this Titmouse are pure white, or grayish-white in ground colour, very sparingly and minutely spotted with pale red, and with underlying markings, similar in character, of gray. Very often the eggs are almost if not entirely devoid of spots, or have the colouring matter suffused indistinctly over the entire surface. Sometimes the spots form a very decided zone. Average measurement, ·55 inch in length, by ·43 inch in breadth. Incubation, performed chiefly by the female, who sits with her long tail resting over her back, sometimes pointing out of the entrance-hole, lasts from twelve to fourteen days.

DIAGNOSTIC CHARACTERS: The exceptionally minute markings, together with their small number, readily distinguish the eggs of this species from those of all others likely to be confused with them. The form and composition of the nest are also an unfailing guide to identification.

Family PARIDÆ. Genus PARUS.
Sub-family PARINÆ.

CRESTED TITMOUSE.

PARUS CRISTATUS, *Linnæus*.

Probably Double Brooded.   Laying season, April to June.

BRITISH BREEDING AREA: The Crested Titmouse is another very local species, being, so far as is known, confined chiefly to the counties watered by the Spey, extending, but more sparingly, into Ross-shire in the north, and to Aberdeenshire in the east.

BREEDING HABITS: The breeding-grounds of the Crested Titmouse are the pine forests, and, less frequently, the oak woods and birch coppices. There can be little doubt that this species pairs for life, as all the winter through it may be seen in company with its mate, and although gregarious to some extent at that season, no social tendency is evinced during the nesting period. It is probable that the same nesting-place is tenanted season by season where the birds are not much disturbed. The nest of the Crested Titmouse is placed in a hole in a tree, a dead branch or trunk being selected by preference, at varying heights from the ground, from a few inches to a dozen feet or more, according to circumstances. If a suitable hole is not accessible, it is said that the birds will make one for themselves in soft, rotten wood, notably in fences. In some extra British localities the bird is said to make its nest in the deserted home of a Crow or a Hawk, or even in the disused drey of a squirrel. Curious sites have been recorded: as, for instance, in boxes placed for the accommodation of Starlings; whilst the bird has even been said to take possession of the nest of a Wren or a Long-tailed Titmouse. The nest is made of moss and dry grass, and lined with wool, hair, fur, and feathers well felted together. It is slovenly made, and varies a good deal in the amount and description of the materials. This Titmouse is a close sitter, usually remaining on the eggs until forcibly removed.

RANGE OF EGG COLOURATION AND MEASUREMENT: The eggs of this species vary from five to eight in number. They are white in ground colour, boldly spotted, blotched, and freckled with brownish-red. The distribution and size of the markings varies considerably. On some eggs they are pretty equally dispersed over the entire surface; in others, they form a zone or even a circular patch round or on the larger end. On some varieties the spots are large, bold, and irregular, and paler in colour than those on other varieties, on which they are small and more or less round in shape. Occasionally the eggs of this species are almost spotless. Average measurement, ·65 inch in length, by ·5 inch in breadth. Incubation, performed by both sexes, lasts from thirteen to fourteen days.

DIAGNOSTIC CHARACTERS: The large size and boldness of the markings on the eggs of the Crested Titmouse are to a certain extent aids to their identification. They however require careful verification. The precise locality is of some service, whilst the conspicuous crests of the parent birds cannot be mistaken.

Family PARIDÆ. Genus PARUS.
Sub-family *PARINÆ*.

MARSH TITMOUSE.

PARUS PALUSTRIS, *Linnæus*, and PARUS PALUSTRIS DRESSERI, *Stejneger*.

Double Brooded.   Laying season, April to June.

BRITISH BREEDING AREA: The Marsh Titmouse is somewhat locally distributed throughout England and Wales, and becomes even more so in Scotland, where it is not known to breed north of the Forth Valley. Its breeding range in Ireland requires definition.

BREEDING HABITS: The breeding-grounds of the Marsh Titmouse are woods, marshy plantations, orchards, hedgerows, and the fringes of alders and pollard willows on the banks of canals and slow-running rivers. Its trivial name, to the uninitiated, implies a swampy habitat, but such is not the case, and this species may be met with in most localities tenanted by birds of the present family. The Marsh Titmouse probably pairs for life, and in some cases, at all events, returns year by year to one particular spot to nest. The nest of the Marsh Titmouse is seldom made more than five or six feet from the ground, and is often not many inches above it. It is usually placed in a hole in a tree or a stump—a rotten stake in the hedgerows, or even a gate-post, is frequently selected, whilst holes in pollard willows and alders are favourite sites, and occasionally a hole in the ground itself, or a crevice in a dry wall, is made use of. In many cases where the wood is soft and rotten the birds excavate a hole for themselves, but where the wood is hard a knot-hole is usually selected. The nest is made at varying depths, sometimes not more than six inches, at others twice or even three times that depth. I have known it rarely as much as a yard. It is a slovenly-made structure of moss, bits of dry grass, wool, feathers, fur, and hair felted together and wedged tightly into the bottom of the hole. The Marsh Titmouse is a close sitter, and commences to hiss and bite when touched by the hand, remaining on the eggs until forcibly removed from them. The parent birds are also very careful in visiting or leaving the nest.

RANGE OF EGG COLOURATION AND MEASUREMENT: The eggs of the Marsh Titmouse are from six to ten in number. They are white in ground colour, spotted and freckled with brownish-red, or reddish-brown. As a rule the spots are most thickly distributed round the larger end of the egg. On some eggs the spots are minute, and scattered over most of the surface, whilst on others they are larger, irregular in shape, and fewer in number. Average measurement, ·62 inch in length, by ·48 inch in breadth. Incubation, performed by both sexes, lasts fourteen days.

DIAGNOSTIC CHARACTERS: It is impossible to give any reliable character by which the eggs of the Marsh Titmouse may be separated from those of allied species. It is absolutely imperative that the parents should be identified, to place the authentication of the eggs beyond doubt.

Family PARIDÆ.   Genus PARUS.
Sub-family PARINÆ.

COAL TITMOUSE.

PARUS ATER, *Linnæus*, and PARUS ATER BRITANNICUS, *Sharpe* and *Dresser*.

Double Brooded. Laying season, April to June.

BRITISH BREEDING AREA: The Coal Titmouse is generally distributed throughout the British Islands, although nowhere so common as the Blue Titmouse. It does not, however, extend its breeding range to the Outer Hebrides, the Orkneys, nor the Shetlands.

BREEDING HABITS: The principal breeding-grounds of the Coal Titmouse are birch and oak coppices, fir plantations, groves of alder trees, pine woods, and large orchards. Less frequently, shrubberies, parks, and old hedgerows are frequented during the nesting period. It is not improbable that the Coal Titmouse pairs for life, and in numbers of instances returns year by year to one particular nest-hole. Certainly the birds may be seen in pairs every month in the year. The Coal Titmouse is more or less gregarious and social during autumn and winter, but all such tendencies lapse during the breeding season, and each pair keep to themselves. The nest of this species is made in a variety of situations, but almost invariably in a hole of some kind, or well sheltered from the external air. The usual site is in a hole of some tree or stump, and less frequently in a crevice or chink in a wall. Occasionally a hole in the ground is selected. In many cases the selected hole is altered and enlarged by the birds. The nest is made at varying depths, sometimes but a few inches, at others a foot or even more. Like that of its congener, the Marsh Titmouse, it is rudely and slovenly made, and is composed of a little dry grass and quantities of moss felted together with hair, and warmly lined with feathers. The birds are careful not to betray the whereabouts of their nest, and are very silent during the period of incubation. The parent bird sits closely, and will hiss and peck at the intruder's hand, remaining brooding over the eggs until forcibly removed.

RANGE OF EGG COLOURATION AND MEASUREMENT: The eggs of the Coal Titmouse are from five to eight, or even ten, in number. They are white in ground colour, spotted and freckled with light red. As a rule most of the markings are distributed over the larger end of the egg, sometimes forming a zone, but frequently they are scattered over most of the surface. On some eggs the markings take the form of minute specks; on others they are larger and more irregular in shape. Average measurement, ·61 inch in length, by ·46 inch in breadth. Incubation, performed by both sexes, lasts fourteen days.

DIAGNOSTIC CHARACTERS: It is impossible to distinguish the eggs of the Coal Titmouse from those of allied species. The only reliable means of identification is to observe the parents—usually a very easy task.

Family PARIDÆ.   Genus PARUS.
Sub-family PARINÆ.

BLUE TITMOUSE.

PARUS CÆRULEUS, *Linnæus*.

Double Brooded.   Laying season, April and June.

BRITISH BREEDING AREA : The Blue Titmouse is by far the commonest species of the present genus found in the British Islands. It is more or less abundantly distributed in all districts suited to its requirements throughout the United Kingdom, but does not, as a breeding species, extend to the Hebrides, the Orkneys, and the Shetlands. It also becomes rarer and more local in the northern parts of Scotland, and the wilder districts of Ireland.

BREEDING HABITS : The Blue Titmouse shows little partiality for any particular haunt. Its breeding-grounds may be said to be anywhere in all districts that are sufficiently well wooded for the requirements of this species. It breeds indiscriminately in woods, fields, lanes, hedgerows, shrubberies, plantations, gardens, orchards, and in buildings of various kinds, as well as walls in suitable localities. I am of opinion that the Blue Titmouse pairs for life, and yearly it will be found to return to its old nest, even in spite of continued disturbance. Although social and gregarious for the remainder of the year, in the breeding season the Blue Titmouse is solitary, each pair keeping to themselves until the young leave the nest. The site for the nest of this species is invariably a covered one. Holes in trees, stumps, posts, and in walls are the favourite situations ; less frequently a hole in the ground is selected. Many curious sites for the nest have been recorded, which we have not space to name, such as pumps, cupboards flower-pots, and the like. The bird will also readily take possession of a box placed in a tree, as I have many times experienced. The nest, made at varying depths, is a slovenly and careless structure of moss and dry grass, felted together with wool and hair, and lined with feathers. The actions of the Blue Titmouse at the nest do not differ in any important respect from those of species already described.

RANGE OF EGG COLOURATION AND MEASUREMENT: The eggs of the Blue Titmouse vary from six to ten in number, but clutches of a dozen are not very uncommon, whilst as many as eighteen have been recorded. They are pure white in ground colour, freckled and minutely spotted with light red. The markings are usually very small, most abundant round the large end of the egg, where they form a zone. Average measurement, ·58 inch in length, by ·45 inch in breadth. Incubation, performed by both parents, lasts fourteen days.

DIAGNOSTIC CHARACTERS: The eggs of the Blue Titmouse so closely resemble those of allied species, that the only safe means of authenticating them is to identify the parents at the nest. As a rule the eggs of this Titmouse are more minutely marked than those of allied species, but the character is not a constant one.

Family PARIDÆ. Genus PARUS.
Sub-family PARINÆ.

GREAT TITMOUSE.

PARUS MAJOR, *Linnæus*.

Double Brooded.   Laying season, April and June.

BRITISH BREEDING AREA: Next to the Blue Titmouse the present species is certainly the commonest and most widely distributed Titmouse found within our area. It breeds more or less abundantly throughout the wooded and well-timbered districts of the British Islands, with the exception of the Hebrides, Orkneys, and Shetlands.

BREEDING HABITS: Confined in a similar manner to the cultivated and timbered districts as the preceding species, its haunts are practically the same. It may be said to breed almost everywhere, amongst trees of all kinds, in woods, plantations, and coppices, in gardens and orchards, in lanes and hedgerows, near houses, and in outbuildings. Like the preceding species it pairs for life, returning year by year to one favourite spot; and although gregarious and social at other times, becomes solitary during the nesting period, each pair keeping to themselves. The birds also become much less noisy as soon as incubation commences, and skulk more closely amongst the foliage. The nest of the Great Titmouse may be conveniently divided into two very distinct types, each made in quite a different situation. First we have the most usual nest in a hole of a tree, post, or stump, or in a crevice of a wall, loosely and carelessly made of felted moss, dead leaves, dry grass, hair, and wool, lined with feathers. Second, we have the much less frequent nest placed in the deserted home of a Crow, Magpie, or squirrel, or amongst the sticks of a Rook's nest, well made, globular, and composed chiefly of moss externally, warmly lined with feathers. Occasionally the nest is made in a hole in the ground, whilst many curious situations have been recorded. Montagu states that a nest is sometimes dispensed with altogether, the eggs being laid on the powdered wood. The bird frequently enlarges the hole, and has been known to excavate it entirely. The actions of the old birds at the nest do not differ from those of their congeners, already described. They are remarkably shy and wary birds throughout the breeding season.

RANGE OF EGG COLOURATION AND MEASUREMENT: The eggs of the Great Titmouse are usually from five to eight in number, but occasionally ten or twelve may be found. They are white or yellowish-white in ground colour, freckled and blotched with light red. As usual, two distinct types of eggs occur: those in which the spots are minute, and those in which they are larger and blotchy; whilst both types either have them uniformly distributed over most of the surface, or confined to a zone round the larger end. Average measurement, ·7 inch in length, by ·55 inch in breadth. Incubation, performed by both sexes, lasts fourteen days.

DIAGNOSTIC CHARACTERS: The eggs of the Great Titmouse are on an average readily distinguished by their larger size from those of allied Titmice, but the eggs of the Nuthatch and the Creeper very closely resemble them; the nest, however, is quite sufficient to designate the ownership.

Family PARIDÆ. Genus REGULUS.
Sub-family REGULINÆ.

GOLDCREST.

REGULUS CRISTATUS, *Koch.*

Single Brooded.   Laying season, April and May.

BRITISH BREEDING AREA: The Goldcrest is generally and widely distributed throughout the British Islands, becoming, of course, the least common in districts where trees are scarce, and therefore not known to breed in the Outer Hebrides, the Orkneys, and the Shetlands. The range of this species is slowly yet steadily increasing in Scotland, owing to the formation of larch and fir plantations.

BREEDING HABITS: The principal breeding-haunts of the Goldcrest are shrubberies and plantations, especially such that contain fir, larch, or yew trees. It is noteworthy what a small plantation will content several pairs of birds, sometimes a mere bunch of Scotch firs or larches containing several nests. It is probable that the Goldcrest pairs for life, although a new nest appears to be made each year. This nest is a charming piece of handiwork, usually made at the extremity of a branch, nearly globular in form, and slung hammock-wise from the twigs. The foliage surrounding it is interwoven with the materials, so that the whole appears nothing but a more than usually dense tuft of vegetation. Nests of this species have been found on the top of flat branches, but this must be very exceptional, as are also nests made in bushes. The nest is composed of moss, spiders' webs, hairs, and a few lichens well felted together, and lined with a warm bed of feathers. The Goldcrest is shy and retiring during the nesting season, and a close sitter, but the peculiar high-pitched callnotes readily betray its presence, and render the discovery of the nest a by no means difficult task.

RANGE OF EGG COLOURATION AND MEASUREMENT: The eggs of the Goldcrest are usually from five to eight in number, but ten have been occasionally found. They are pale reddish-white in ground colour, sometimes pure white, minutely speckled with brownish-red, chiefly on the larger end. Some eggs have the colouring matter confluent, clouded, and suffused over the entire surface. Average measurement, ·53 inch in length, by ·41 inch in breadth. Incubation, performed chiefly by the female, lasts twelve or thirteen days.

DIAGNOSTIC CHARACTERS: The small size, minute markings, and reddish tinge are sufficient to distinguish the eggs of the Goldcrest from those of any other species breeding in the British Islands. The nest also is unique.

Family LANIIDÆ.   Genus LANIUS.

RED-BACKED SHRIKE.

LANIUS COLLURIO, *Linnæus*.

Single Brooded.   Laying season, May and June.

BRITISH BREEDING AREA: The Red-backed Shrike is decidedly a southern species, but one nevertheless pretty generally distributed during summer over the southern and central portions of England and Wales. It is said to be rare in Cornwall, but is certainly common in Devonshire, and thence in all the southern counties to Kent. In Norfolk and Lincolnshire it is said to be decreasing, and anywhere north of Yorkshire is decidedly rare and local. It has occasionally bred in the south-east of Scotland, but is unknown as a nesting species in Ireland.

BREEDING HABITS: The Red-backed Shrike is one of the latest of our summer visitors to arrive. The haunts of this species are farms where tall hedges are numerous, common lands, covered with plenty of thicket and brushwood, country lanes, orchards, and large gardens. I am of opinion that this bird pairs for life. It certainly migrates in pairs, and appears to return with wonderful regularity to its old haunts, even the same bushes being visited year by year. It seldom wanders far from one locality from the time of its arrival early in May, until the young are able to fly, in July or August. The nest is built in a variety of situations, generally with no attempt at concealment, and very often by the side of the highway or near a footpath. A favourite spot is in a tall hedge, or a thick bush, less frequently in a thicket of briars and brambles. One spot, visited yearly by a pair of these Shrikes, was a large clump of willows on the Great Western Railway embankment between Paignton and Torquay, and not a dozen feet from the passing trains. It has now been cut down, and the birds have gone elsewhere. The nest is built at varying heights, on an average about six or eight feet from the ground. It differs a good deal, not only in the materials, but in the amount of care and skill displayed in its construction. The usual type of nest is made externally of round dry grass-stems, the stalks of various plants, and a few roots; internally, it is composed of finer stems and roots, a few feathers and flakes of moss and wool (round the rim), and finally lined with hair. A less frequent type of nest now lying beside me is composed chiefly of moss and wool felted together, and strengthened on the outside with round dry grass-stalks, and a few dead twigs (oak, woodbine, wild-rose, and bramble). The nest is very loosely built into the site, sometimes a few of the surrounding twigs being enclosed with the outer materials. The male makes himself very conspicuous near the nest, and his actions, if observed, soon betray its whereabouts. The female sits closely, and makes little demonstration when disturbed from her eggs. I have known this bird make several nests one after the other in a certain spot as they have been taken.

RANGE OF EGG COLOURATION AND MEASUREMENT: The eggs of the Red-backed Shrike are from four to six in number, the former often being found. They vary considerably in colour, but so far as I can determine each bird constantly lays one certain type. The ground colour is either pale green, pale buff, white, creamy-white, or salmon colour. The green and buff varieties are usually spotted, freckled, and blotched with various shades of olive-brown, and the underlying markings, similar in character, are pale brown and violet-gray. The white, cream, and salmon-tinted varieties are usually spotted, freckled, and blotched with reddish-brown and brownish-red, and with underlying markings of violet-gray. The markings are generally most numerous on the larger end of the egg, where they form a broad, irregular zone, but sometimes they are more evenly distributed. The pale underlying markings are both numerous and well defined; indeed on some eggs they predominate. Occasionally, a few streaks may be met with. Average measurement, ·88 inch in length, by ·65 inch in breadth. Incubation is performed almost exclusively by the female, and lasts fourteen days.

DIAGNOSTIC CHARACTERS: The eggs of this Shrike are not readily confused with those of any other species habitually breeding within the British area, their size, well-marked types of colour, and zones of spots being so far peculiar to them as to render their identification an easy matter. The nest is also very characteristic.

The Woodchat Shrike (*Lanius rufus*) has been recorded as having bred several times in the Isle of Wight, but as our islands are somewhat out of the normal line of migration of this species, the event must be an exceedingly exceptional one. I had abundant opportunities of studying the habits of this handsome bird in Algeria. The eggs are usually laid in May. The nest is made in a fork of the branches of a small tree or large bush; I have seen it twenty feet from the ground, close to the trunk of a poplar; and in many cases it is made with little or no attempt at concealment. Externally the materials are stalks and stems of plants and grass, the dry leaves of grass, dead leaves, and occasionally tufts of wool; it is lined with withered flowers of the cudweed and other aromatic plants, wool and vegetable down. The eggs are from four to six in number, five being the usual clutch. They are considerably larger than those of the Red-backed Shrike, but otherwise closely resemble them in appearance. It may be remarked, however, that the red type of egg so common in that species is just as rare in the eggs of the Woodchat Shrike. Average measurement, ·92 inch in length, by ·69 inch in breadth. Incubation, performed almost entirely by the female, lasts fourteen days. But one brood appears to be reared in the year.

Family TURDIDÆ. Genus PHYLLOSCOPUS.
Sub-family SYLVIINÆ.

CHIFFCHAFF.

PHYLLOSCOPUS RUFUS (*Bechstein*).

Single Brooded. Laying season, April, May, and June.

BRITISH BREEDING AREA: The Chiffchaff is fairly well distributed over England and Wales, most abundant in the southern and midland counties, becoming rarer and more local northwards. In Scotland and the north of England it is nothing near so common as the Willow Wren, but certainly appears to extend as far as Ross-shire. Its breeding area in Ireland is very imperfectly known, but the bird is certainly more local than in England.

BREEDING HABITS: The Chiffchaff is a summer migrant, reaching our shores in March and April. The haunts of this species are woods, plantations, and coppices, tall hedgerows and orchards, whence throughout the spring the monotonous song or *chiff-chaff* of the male is almost incessantly sounding. The Chiffchaff may pair annually, although there is some evidence to suggest that certain spots are visited by certain birds each season. The nest is either built on the ground or, more generally, from a few inches to a few feet above it. It is placed amongst tall, rank vegetation, growing in the woods or by the hedge-side, in ivy, either growing up a tree or wall, and least frequently of all in a bush or mass of brambles. The nest of the Chiffchaff is oval and semi-domed, the hole admitting the parents being on the side near the top. It is made externally of dry grass, dead leaves, and scraps of moss, and lined with a little horsehair and even roots, and a large quantity of feathers. The Chiffchaff is a close sitter, and when disturbed glides very quietly from her eggs. In visiting the nest the old birds are exceedingly cautious, if any danger threatens, and hop restlessly about for a long time before they will betray their secret. This species is not at all social during the nesting season.

RANGE OF EGG COLOURATION AND MEASUREMENT: The eggs of the Chiffchaff are from five to seven in number. They range from white to white tinged with yellow in ground colour, somewhat sparingly spotted and freckled with dark or pale reddish-brown, and with a few underlying markings of violet-gray. Most of the spots are, as a rule, on the larger end of the egg, but occasionally they are more uniformly distributed. Average measurement, ·6 inch in length by ·46 inch in breadth. Incubation, performed chiefly by the female, lasts thirteen or fourteen days.

DIAGNOSTIC CHARACTERS: The eggs of the Chiffchaff are best distinguished by their few and dark reddish-brown coloured markings. The eggs of some of the Titmice closely resemble them, but the formation and locality of the nest prevent confusion.

Family TURDIDÆ.　　　　　Genus PHYLLOSCOPUS.
Sub-family *SYLVIINÆ.*

WILLOW WREN.

PHYLLOSCOPUS TROCHILUS (*Linnæus*).

Single Brooded. Laying season, April, May, and June.

BRITISH BREEDING AREA: The Willow Wren is commonly distributed throughout the British Islands, extending to almost every portion where trees occur, although it does not breed in the Orkneys and Shetlands.

BREEDING HABITS: The haunts of the Willow Wren embrace almost every kind of scenery, provided trees or bushes form a part. The bird may be found breeding near the moors, in the coppices and plantations of young trees, or amongst the trees and bushes that fringe the mountain stream. In better-cultivated districts it may be met with amongst the fields and orchards and gardens, in shrubberies, woods, and game coverts—in fact wherever trees with a moderate amount of undergrowth below them are to be found, the sweet song and plaintive call-note of the Willow Wren may be heard. It is undoubtedly the most common of all the Warblers that visit us in spring, arriving early in April. Whether it pairs for life or not is difficult to say; but I think the same spots are visited yearly, although the birds do not join into pairs for some time after their arrival. The nest is usually made on the ground, among herbage on a bank, and under the shelter of a bush or tuft of tall grass by preference. Frequently it is made among mowing grass, and very rarely it may be met with at some distance from the ground. I once found a nest some three feet from the ground, resting on a stone jutting from an old wall, surrounded with ivy, and partly supported by the stem of a hawthorn sapling. The nest is semi-domed, but more open than that of the Chiffchaff, loosely put together, and made externally of dry grass, scraps of moss, withered leaves, and roots, and lined with horsehair, cowhair, and quantities of feathers. I have counted two hundred feathers in the lining of a single nest! The parent birds are very wary at the nest, but may generally be made to disclose its whereabouts if sufficient patience is used. The bird is a close sitter, and when flushed will flit restlessly about from spray to spray near the cleverly-concealed nest, not venturing to visit it until all fear is allayed.

RANGE OF EGG COLOURATION AND MEASUREMENT: The eggs of the Willow Wren are from four to seven or even eight in number. They range from white to white tinged with yellow in ground colour, blotched, spotted, and freckled with pale brownish-red or reddish-brown. The markings are frequently distributed over most of the surface of the shell, occasionally most numerous on the larger end, and forming a zone or circular, semi-confluent patch. The amount of spotting varies considerably. Average measurement, ·62 inch in length, by ·47 inch in breadth. Incubation, performed by both sexes, lasts from thirteen to fourteen days.

DIAGNOSTIC CHARACTERS: The *pale* reddish-brown or brownish-red markings, combined with the locality and formation of the nest, serve to distinguish the eggs of the Willow Wren, and to render their identification easy.

Family TURDIDÆ.   Genus PHYLLOSCOPUS.
Sub-family SYLVIINÆ.

WOOD WREN.

PHYLLOSCOPUS SIBILATRIX (*Bechstein*).

Single Brooded. Laying season, May and June.

BRITISH BREEDING AREA: The Wood Wren is by far the rarest and most local of the three British Willow Warblers. It is more or less thinly dispersed throughout suitable districts in England and Wales; and becomes even more sparingly distributed in Scotland, although certainly breeding as far north as Sutherlandshire. In Ireland it is rare, but probably breeds in Wicklow and some other counties on the eastern seaboard.

BREEDING HABITS: The favourite breeding-grounds and summer haunts of the Wood Wren are large woods and coppices, plantations of tall mixed trees with plenty of undergrowth, game coverts, and, less frequently, orchards. It is a late migrant, not reaching the British Islands before the middle or end of April. It is impossible to say whether this species pairs for life; the sexes do not appear to migrate in company. I have observed a tendency to frequent a certain spot year by year. The nest is invariably placed on the ground often wedged in a little hollow bare of herbage. A bank in the woods well clothed with rank vegetation, or a more open spot in the coppices and plantations amongst heath and bilberry wires, are the favourite situations. The nest, in shape similar to that of the Willow Wren or the Chiffchaff, semi-domed, is made principally of dry grass, with scraps of moss and a few dead leaves, and lined sparingly with horsehair. The Wood Wren is another close sitter. The nest is cunningly concealed, and usually found as the bird hurriedly flies off. When disturbed, the actions of the parents are very similar to those of the Willow Wren, and the nest may usually be discovered with due patience. The Wood Wren is not at all social during the nesting season. The monotonous double call-note (*dee-ur*), and the strange shivering song are ready means of identifying this species.

RANGE OF EGG COLOURATION AND MEASUREMENT: The eggs of the Wood Wren are from five to seven in number. They are pure white in ground colour, thickly spotted and freckled with rich purplish-brown, and with underlying markings of violet-gray. Usually the markings are pretty evenly distributed, becoming most numerous round the larger end of the egg, often forming a zone. Some are much more thickly marked than others; and on some many of the markings run into large pale blotches intermingled with smaller and darker spots. The underlying markings are both well defined and numerous. Average measurement, ·65 inch in length, by ·56 inch in breadth. Incubation, performed chiefly by the female, lasts from thirteen to fourteen days.

DIAGNOSTIC CHARACTERS: The numerous and rich dark brown surface-markings, and violet-gray underlying spots, readily distinguish the eggs of the Wood Wren from those of allied species breeding in our islands. The absence of feathers in the nest-lining is also another important fact in the question of their identification.

Family TURDIDÆ. Genus SYLVIA.
Sub-family *SYLVIINÆ*.

DARTFORD WARBLER.

SYLVIA PROVINCIALIS (*Gmelin*).

Double Brooded. Laying season, April and June.

BRITISH BREEDING AREA: The Dartford Warbler is another very local species, and with a somewhat restricted area of distribution in our islands. It breeds locally from Cornwall to Kent, thence northwards along the Thames valley and through some of the midland districts

(Worcestershire, Leicestershire, Derbyshire) south of Yorkshire. A few also probably breed in Cambridgeshire, Norfolk, and Suffolk. I have taken the nest of this bird from a gorse covert by the side of the reservoirs at Hollow Meadows, on the extreme southern borders of Yorkshire. Unknown in Ireland and Scotland.

BREEDING HABITS: The favourite and, during the breeding season, so far as our islands are concerned, exclusive haunts of the Dartford Warbler are gorse or furze coverts. It is a resident species, and does not wander far from its usual retreats all through the year. It is difficult to say whether this species pairs for life or not; certainly the birds keep close company always, and may be found nesting season after season in certain favoured places. The Dartford Warbler is very skulking in its habits, and easily overlooked, although the peculiar and very characteristic note of *pit-it-chou*, or the more Whitethroat-like scolding *chay-chay-chay*, often reveals its presence when the bird is hidden from view amongst the dense and well-nigh impenetrable cover. The nest of this Warbler is almost invariably built amongst the withered lower branches of the furze, where very often long dry grass is matted round the stems. It is a very delicate and loosely-made structure, composed externally of round dry grass-stalks, bits of withered furze and scraps of moss, and lined with flakes of wool, finer grass-stems, and occasionally a few horsehairs. It is somewhat deep, but very net-like and fragile. The parent birds are very skulking at the nest, the hen sitting closely, and when disturbed slipping quietly off into the dense surrounding cover. No social tendencies are manifest in the breeding habits of this species.

RANGE OF EGG COLOURATION AND MEASUREMENT: The eggs of the Dartford Warbler are four or five in number. They are white in ground colour, sometimes suffused with green or buff, spotted and freckled with dark brown, and with underlying markings of paler brown and gray. The spots are usually most numerous on the larger end of the egg, often forming an irregular zone, but sometimes they are more evenly dispersed over the entire surface. Average measurement, ·68 inch in length, by ·51 inch in breadth. Incubation, performed chiefly by the female, lasts from twelve to fourteen days.

DIAGNOSTIC CHARACTERS: The eggs of the Dartford Warbler very closely resemble those of the Whitethroat, but as a rule the markings are darker and more clearly defined. The eggs of this Warbler require the most careful identification. The nest in some measure assists the collector in the task.

Family TURDIDÆ. Genus SYLVIA.
Sub-family *SYLVIINÆ*.

LESSER WHITETHROAT.

SYLVIA CURRUCA (*Linnæus*).

Single Brooded. Laying season, May and June.

BRITISH BREEDING AREA: The Lesser Whitethroat is a much more local species than the Common Whitethroat, becoming rare in the west of England and in Wales. It is fairly well distributed over the southern, central, and eastern counties, as far north as Lancashire and Yorkshire. It becomes rare north of these limits; whilst in Scotland it is even rarer and more locally dispersed, as far north as the valley of the Forth. Of only abnormal occurrence in Ireland (*Zoologist*, 1891, p. 186).

BREEDING HABITS: The breeding-grounds of the Lesser Whitethroat are coppices, small plantations, orchards, gardens, commons, hedgerows, and shrubberies. It is a summer migrant to our islands, arriving during the latter half of April or early in May. It appears to pair annually, shortly after reaching its nesting localities. A peculiarity of its habits is its partiality for the higher branches of the trees and hedges. The nest is placed amongst the dense lower vegetation of its haunts, but occasionally in the topmost branch of a tall hedge or thicket. Frequently it may be found amongst brambles and briars, in gorse bushes, and in branches overhanging a dell or a stream. It is a shallow, slight, net-like structure, made of dry grass-stalks, often bound together with cobwebs or cocoons and roots, and lined with a little horsehair. The hen-bird sits very closely, and when disturbed flits to and fro from branch to branch, often in company with the male, from time to time uttering a harsh and scolding *tec* or *tay-tay*. This species is not at all social during the nesting season.

RANGE OF EGG COLOURATION AND MEASUREMENT: The eggs of the Lesser Whitethroat are four or five in number. They vary from pure white to very pale buff in ground colour, blotched and freckled with various shades of greenish-brown, and with underlying markings of violet-gray. On some eggs a few dark brown streaks occur. The markings are usually most abundant on the large end of the egg, where they form a zone or a semi-confluent circular patch. On some eggs most of the markings are large and pale, and most of the larger spots are paler round the margin. Average measurement, ·65 inch in length, by ·52 inch in breadth.

Incubation, performed chiefly by the female, lasts from twelve to fourteen days.

DIAGNOSTIC CHARACTERS: The clearly-defined spots and strongly-contrasting ground colour, combined with the size, distinguish the eggs of the Lesser Whitethroat from those of species likely to be confused with them. The nest also is characteristic.

Family TURDIDÆ. Genus SYLVIA.
Sub-family *SYLVIINÆ*.

WHITETHROAT.

SYLVIA CINEREA (*Bechstein*).

Single Brooded. Laying season, May and June.

BRITISH BREEDING AREA: The Whitethroat is one of the commonest and most widely distributed of the Warblers. It is found during the breeding season in all suitable localities throughout England, Wales, and Ireland; whilst in Scotland, although a little more local, it breeds regularly at least as far north as Ross-shire, including Mull, Iona, and Skye.

BREEDING HABITS: The Whitethroat arrives in our islands at the end of April or early in May. Its favourite haunts are tangled hedgerows and thickets, both in cultivated districts and on the moorlands. The males generally arrive in their summer quarters a short time before the females, after whose appearance pairing becomes general. The nest may be found in a great variety of situations, sometimes within a few inches of the ground, more generally a few feet, and occasionally at the top of a tall hedge, six or eight feet high. It is a flimsy, net-like structure, often placed amongst nettles and other coarse herbage, amongst briars and brambles, or in a dense whitethorn, woodbine, or other bush. It is very deep for its size, beautifully rounded, and formed almost entirely of dry grass-stalks and a few roots, cemented with one or two cobwebs and cocoons, and lined with horsehair. Both male and female assist in making the nest, and from what I have repeatedly observed I should say the cock does the greater part of the work. When disturbed from the nest the female slips quietly off into the cover, with perhaps a warning *tec*, but the male is more demonstrative, and scolds and puffs out his plumage in alarmed annoyance. This Warbler is not at all social during the nesting period.

RANGE OF EGG COLOURATION AND MEASUREMENT: The eggs of the Common Whitethroat are from four to six in number. They vary in ground colour from pale bluish-white to buffish-white and pale green, mottled, blotched, and freckled with light brown and olive-brown, and with underlying markings of pale pearly-brown and violet-gray. On some eggs most of the markings are underlying ones, and gray in tint; whilst rarely spots of dark brown occur. The distribution of the spots varies in the usual manner; on some eggs they are evenly dispersed, on others they form a strongly-marked zone, or even a circular patch, palest round the margin. Average measurement, ·72 inch in length, by ·55 inch in breadth. Incubation, performed chiefly by the female, lasts from eleven to thirteen days.

DIAGNOSTIC CHARACTERS: Certain eggs of the Whitethroat very closely resemble those of the Dartford Warbler, but usually their clouded green or buff appearance distinguishes them from those of the rarer bird; whilst the structure of the nest prevents confusion.

PLATE III.

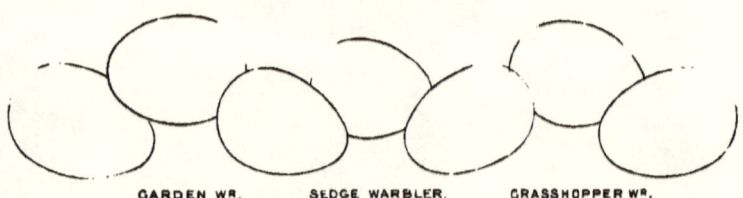

BLACKCAP WR.   GARDEN WR.   SEDGE WARBLER.   GRASSHOPPER WR.
              MARSH WR.   REED WR.   SAVI'S WR.

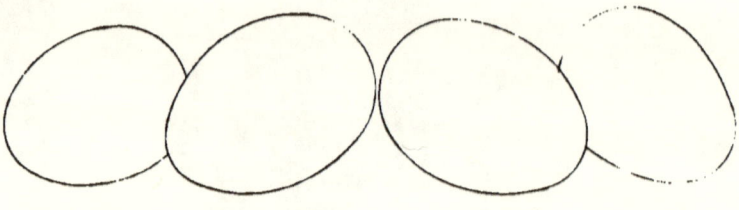

SONG THRUSH   MISSEL THRUSH.   BLACKBIRD.   RING OUZEL.

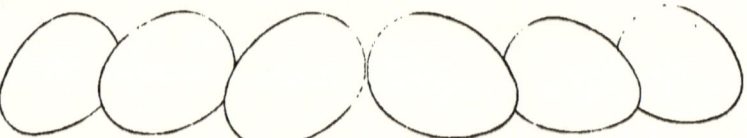

REDSTART.   ROBIN   NIGHTINGALE.   WHEATEAR.   WHINCHAT.   STONECHAT.

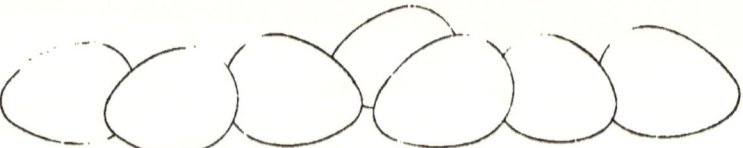

COM WREN   ST KILDA WREN   ACCENTOR.   SP. FLYCATCHER.   P. FLYCATCHER.   SWALLOW.

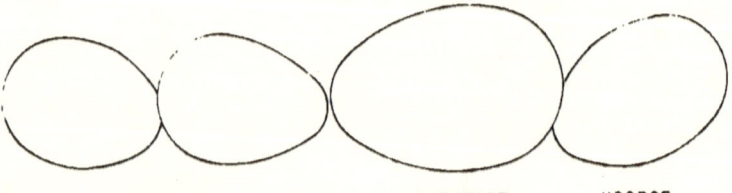

CUCKOO.   NIGHTJAR   HOOPOE

Family TURDIDÆ. Genus SYLVIA.
Sub-family SYLVIINÆ.

GARDEN WARBLER.

SYLVIA HORTENSIS (*Gmelin*).

Single Brooded. Laying season, May and June.

BRITISH BREEDING AREA: The Garden Warbler is generally but locally distributed in all parts of England, with the exception of West Cornwall; in Wales it is not known to breed except in Pembrokeshire and Breconshire. In Scotland it breeds locally in the southern and central districts, probably as far north as Banffshire. In Ireland it is decidedly rare, but breeds in counties Cork, Tipperary, Fermanagh, and Antrim, and may do so in those of Dublin and Wicklow.

BREEDING HABITS: The Garden Warbler is a late migrant, not reaching the British Islands before the beginning of May. Its favourite haunts are thickets and underwood in coverts, plantations, and shrubberies. It is also fond of frequenting the wooded banks of streams, and may often be met with in large gardens, orchards, and nurseries. The males arrive a little before the females. As soon as the latter appear the birds begin to pair, and nest-building commences shortly afterwards. The nest is generally placed not many feet above the ground, amongst brambles, briars, or low thorn trees, and frequently in gooseberry and currant bushes, or even amongst nettles or growing peas. It is a flimsy, net-like structure, made of fine dry grass-stalks, amongst which, in the foundation, a flake or two of moss or a few roots are placed, and lined with a little horsehair. The nest is woven round the surrounding twigs, and is very neatly finished and exquisitely rounded inside. The parents are shy and retiring birds, close sitters, and slipping off quietly into the cover when disturbed, where a trembling twig or the harsh call-note of *tec* is the only sign of their presence.

RANGE OF EGG COLOURATION AND MEASUREMENT: The eggs of the Garden Warbler are usually four or five in number, sometimes as many as six. They vary in ground colour from pure white to white tinged with buff or green, blotched, spotted, mottled, and freckled with olive-brown, dark brown, and buffish-brown, and with underlying markings of paler brown and gray. Some eggs are handsomely blotched and clouded, the larger and paler markings being intermingled with smaller and darker spots; others have the small dark spots predominant and clustering round the larger end, mingled with gray underlying ones; whilst in others most of the markings are underlying ones—large gray blotches, with a few darker surface-spots between them. Average measurement, ·75 inch in length, by ·6 inch in breadth. Incubation, performed chiefly by the female, lasts fourteen days.

DIAGNOSTIC CHARACTERS: The eggs of the Garden Warbler cannot always be distinguished from those of the Blackcap, although the rufous type does not appear to occur. They require the most careful identification. It may be remarked that the male has no black cap, neither has the female the chestnut cap which distinguishes that sex of the Blackcap.

Family TURDIDÆ. Genus SYLVIA.
Sub-family SYLVIINÆ.

BLACKCAP WARBLER.

SYLVIA ATRICAPILLA (*Linnæus*).

Single Brooded. Laying season, May and June.

BRITISH BREEDING AREA: The Blackcap is generally, although somewhat locally, distributed throughout England and Wales. In Scotland it becomes rarer and even more local, especially so north of the valley of the Forth, although it has been met with breeding as high as Ross-shire. In Ireland it is even more local and sparingly dispersed, but certainly breeds in counties Dublin, Wicklow, Tipperary, and Mayo. Further observation may show that it is much more generally distributed.

BREEDING HABITS: The Blackcap arrives in the British Islands about the middle of April. Its principal breeding-haunts are plantations, spinneys, shrubberies, the banks of streams clothed with dense underwood, hedgerows, orchards, and gardens. The Blackcap may be seen in pairs a week or so after its arrival, and nest-building commences soon after that event. The nest is variously placed among briars and brambles, in holly trees, or in thickets growing over the stream, and in dense hedges, sometimes only a foot or so from the ground, at others as much as ten or twelve feet. Both birds assist in its construction. It is a rather slight, flimsy, and loosely-woven structure, composed principally of dry grass-stalks, perhaps a scrap or two of moss, a few leaf-stalks and roots, and sparingly lined with horse-hair. Many nests have cobwebs and cocoons amongst them. Blackcaps are close sitters, often staying on the eggs until the bush is shaken. The sitting bird then glides quietly off into the cover, but will frequently hop restlessly about at arm's length, occasionally uttering the harsh and warning *tcc*. The cock-bird occasionally sings whilst sitting on the nest. Nest after nest will sometimes be made in one locality, as often as they are removed.

RANGE OF EGG COLOURATION AND MEASUREMENT: The eggs of the Blackcap are four or five in number, sometimes six; and in cases where the first clutch has been taken, only three. Three distinct types occur in the eggs of this species. The most frequent of these is dull white in ground colour, clouded and suffused with olive or buffish-brown, and here and there spotted and occasionally streaked with rich dark brown. The second type has the ground colour suffused with very pale blue, blotched and spotted with yellowish-brown, and with a few streaks of darker brown, and with numerous and large underlying markings of gray. The rarest type is pale brick red in ground colour, marbled and clouded with darker red, and sparingly spotted and streaked with dark purplish-brown. Average measurement, ·8 inch in length, by ·56 inch in breadth. Incubation, performed almost equally by both sexes, lasts fourteen days.

DIAGNOSTIC CHARACTERS: Many eggs of the Blackcap are undistinguishable from those of the Garden Warbler, but the red type seems peculiar to the present species. They require very careful identification.

Family TURDIDÆ. Genus ACROCEPHALUS.
Sub-family *SYLVIINÆ.*

MARSH WARBLER.

ACROCEPHALUS PALUSTRIS (*Bechstein*).

Single Brooded. Laying season, June.

BRITISH BREEDING AREA: The Marsh Warbler is undoubtedly the most local of all the Warblers that visit our islands to breed. It is a rare and local visitor to Somersetshire, more especially to the vicinity of Bath and Taunton. As its nest, however, has been taken in Cambridgeshire, and the bird itself obtained in Gloucestershire, it is only reasonable to presume that further research will lead to an important extension of its area in our islands. There can be no doubt that the present species has been and continues to be confused with the nearly allied Reed Warbler. Fortunately, however, the eggs of the two species cannot be confused.

BREEDING HABITS: The Marsh Warbler is another late bird of passage, not arriving in our islands before May. But little of its habits in this country has been observed. Its haunts are swamps and osier beds, where plenty of brushwood and rank undergrowth occur. In this respect it more closely resembles the Sedge Warbler than its much nearer ally the Reed Warbler. It also appears partial to the swampy sides of hedges, where a rank and luxuriant vegetation flourishes. It is not a bird of the reeds, always preferring swampy ground to water. The bird appears to pair shortly after its arrival, but nest-building does not commence usually for several weeks after that event. The nest is generally placed (in this country) amongst rank vegetation, not far from the ground. One of the two nests of this bird, taken near Taunton, which I had the pleasure of examining some years ago, was built round stems of the greater willow herb and nettles, about twelve inches from the ground; the other was about three feet from the ground, and supported by stems of figwort and nettles. The nests of the Marsh Warbler, although suspended, as is the case with that of the Reed Warbler, are never placed over water, and sometimes at a considerable distance from open water of any kind. The two nests in question were made almost completely of round, dry grass-stalks, with a few dry grass-leaves and bits of downy fibre and moss intermixed, one sparingly, the other profusely lined with black horsehair, coming right up to the rim. They are rather deep, the cup being beautifully rounded and finished. If the first nest is taken, another will soon be made, and this has been repeated three times in succession. During the nesting season the parent birds are remarkably wary and shy, the sitting bird slipping quietly off the eggs when disturbed, and hiding in the surrounding vegetation. Although not social during the breeding season several nests may be found within a comparatively small area of suitable ground.

RANGE OF EGG COLOURATION AND MEASUREMENT: The eggs of the Marsh Warbler are from four to seven in number, and may be divided into several well-marked types. One type has a greenish-blue ground colour, sparingly spotted and blotched with olive-brown, and with numerous underlying markings of violet-gray; on this type most of the blotches have a darker brown central spot upon them. Another is paler (greenish-white) in ground colour, heavily mottled, clouded, blotched, and spotted with olive-brown, intermingled with a few dark specks, and with the underlying markings few and small. On both types most of the markings are distributed on the larger end of the egg. Average measurement, ·73 inch in length, by ·55 inch in breadth. Incubation, performed chiefly by the female, lasts fourteen days.

DIAGNOSTIC CHARACTERS: The much clearer and whiter ground colour of the eggs of the Marsh Warbler distinguish them from those of the closely-allied Reed Warbler. The situation and construction of the nest are also different.

Family TURDIDÆ. Genus ACROCEPHALUS.
Sub-family SYLVIINÆ.

REED WARBLER.

ACROCEPHALUS ARUNDINACEUS (*Brisson*).

Single Brooded. Laying season, May and June.

BRITISH BREEDING AREA: The Reed Warbler is a summer visitor to our islands, reaching them towards the end of April or early in May. It is somewhat locally distributed, being rare or entirely absent from many apparently suitable localities, and abundant in others. Its principal head-quarters are in the southern, midland, and eastern counties. I do not think it breeds anywhere on the south-western peninsula of England, including Devon and Cornwall. In Wales it is rare and local, but breeds in Breconshire, and is said to be common near the Llangorse Lake. It becomes rare and local in Lancashire and Cumberland, and is only known to breed as far north as Ripon in Yorkshire, but does so in considerable numbers at Hornsea Mere. It has also been said to breed on the Durham side of the Tyne.

In Scotland no reliable instances of its breeding are known; whilst its summer range does not extend to Ireland.

BREEDING HABITS: The principal summer haunts of the Reed Warbler are large reed beds, and the fringe of that vegetation which skirts the banks of canals, slow-running rivers, and dykes. It is also to be met with in osier beds, clumps of willow and alder trees, and brushwood of all kinds growing near the water. Instances are on record of this species even frequenting gardens. The Reed Warbler is not gregarious nor even social, but numbers of pairs may be met with in a very small area, each, however, keeping to a particular spot, from which they jealously drive intruders. The birds pair soon after their arrival, the various nesting-places being selected apparently with much noisy song and vigorous scolding, but the period of nesting depends a good deal on the state of the weather. If the season is warm and early, nest-building commences in May; if cold and backward, not before the beginning of June. The nest is either made amongst the reeds, or in the slender branches of willows and alders overhanging the water. Occasionally it has been met with in hedges by the water-side. It is built at varying heights from the ground, or rather water, often not more than six or eight inches from it, sometimes as low as three feet, at others as much as six, ten, or twelve feet above it. The nests in reeds are suspended between three or four stems, whilst those in thickets are slung from several slender twigs, the materials being woven round each in turn. They are deep, well-made structures, composed principally of dry sedgy grass, broad dead leaves of the reeds, and roots, and lined with finer roots, sometimes a feather or two, and a little hair. Some nests often contain a little moss, bits of wool, and vegetable down. In many cases the nests are very elongated, almost funnel-shaped, and have a considerable foundation for the cup which contains the eggs ; but in others the materials are less in quantity, and this lower structure is dispensed with. When disturbed the parent birds become very anxious and noisy, flitting from stem to stem uttering harsh, scolding notes. Very often the nest will be left for a day or so when finished before an egg is laid in it. If the first nests and eggs are taken, they are almost invariably replaced.

RANGE OF EGG COLOURATION AND MEASUREMENT: The eggs of the Reed Warbler are four or five in number. They are very pale greenish-blue in ground colour, spotted and blotched with greenish-brown, freckled and sometimes streaked with darker brown, and with underlying markings of gray or paler brown. On some varieties the markings are large and more or less confluent, especially at the wide end of the egg ; on others the spots are smaller and more evenly dispersed ; on some the underlying markings are most numerous, on others they are few and indistinct. Average measurement, ·73 inch in length, by ·53 inch in breadth. Incubation, performed chiefly by the female, lasts thirteen or fourteen days.

DIAGNOSTIC CHARACTERS : The darker ground colour readily distinguishes the eggs of the present species from those of the Marsh Warbler, the only ones with which they are likely to be confused. The construction and situation of the nest are also important guides to their correct identification.

Family TURDIDÆ. Genus ACROCEPHALUS.
Sub-family SYLVIINÆ.

SEDGE WARBLER.
ACROCEPHALUS PHRAGMITIS (*Bechstein*).

Single Brooded.   Laying season, May and June.

BRITISH BREEDING AREA: The Sedge Warbler is commonly distributed in all suitable localities throughout the British Islands, with the exception of the Hebrides (it breeds however in Skye), possibly the Orkneys, and certainly the Shetlands, although it becomes rather more local and less abundant in the extreme northern districts.

BREEDING HABITS: The Sedge Warbler arrives in the British Islands near the end of April or the beginning of May. Its haunts are by no means always of a marshy nature, but the bird is certainly most abundant in osier beds, reed beds where the ground is more swampy than full of open water, and the tangled thickets and marshy spinneys near the water-side. Less frequently it may be seen in dense hedgerows at some distance from water, but much more commonly in hedges that border a stream. Like the Reed Warbler the Sedge Warbler is often very abundant in a favourite haunt, and directly after its arrival is very quarrelsome and noisy until all have taken possession of certain spots, to which they keep pretty closely through the nesting season. The pertinacity with which the cock-birds sing, day and night, readily proclaims the presence of this species in a district, and as the bird is far from shy, and hops about the vegetation in full view, its identification is a matter of small difficulty. The birds pair soon after arrival, and nest-building begins almost at once. The nest is variously placed in willow bushes, amongst dense hedges and thickets of bramble and briar, and in the long rank grass and various other weeds that grow round the stumps of the osiers; frequently it is placed on the top of one of the stumps, concealed by surrounding grass and the shoots of the willows. A nest has been recorded built in a gooseberry bush. It is a small and loosely-made structure, and placed either on the ground itself, or from a few inches to ten or fifteen feet above it. It is made of coarse dry grass, bits of moss and withered sedge, and lined with a little horsehair, a scrap or two of vegetable down, and more rarely a few feathers. The nest is difficult to find; the parent bird sits closely, and whenever possible glides off very quietly into the nearest cover.

RANGE OF EGG COLOURATION AND MEASUREMENT: The eggs of the Sedge Warbler are five or six in number, and may be divided into two well-marked types, between which extremes almost every intermediate variety may be obtained. The ground colour of all eggs is bluish-white. The first type is clouded with pale buff, and indistinctly mottled with yellowish-brown; the second is also washed with pale buff, and the markings are bolder in definition and a much richer brown in colour. Almost invariably a few scratches or streaks of blackish-brown occur, chiefly on or near the large end of the egg. Average measurement, ·68 inch in length, by ·52 inch in breadth. Incubation, performed mostly by the female, lasts fourteen or fifteen days.

DIAGNOSTIC CHARACTERS: The clouded buff or yellowish-brown appearance of the eggs of the Sedge Warbler, together with the black lines or pencillings, readily distinguish them from those of allied species.

Family TURDIDÆ.   Genus LOCUSTELLA.
Sub-family SYLVIINÆ.

GRASSHOPPER WARBLER.

LOCUSTELLA LOCUSTELLA (*Latham*).

Single Brooded.   Laying season, May and June.

BRITISH BREEDING AREA : The Grasshopper Warbler is very widely distributed throughout England and Wales, breeding more or less commonly in every county. In Scotland it becomes rarer and more local, chiefly being found in the counties south of the Forth valley; but it breeds sparingly here and there as far north as Skye. In Ireland it is also widely distributed but somewhat local, and chiefly confined to the eastern and southern counties.

BREEDING HABITS : The Grasshopper Warbler arrives in our islands towards the end of April or early in May. Its favourite haunts are amongst tangled thickets and brushwood, either in plantations, spinneys, or coppices, in fens, or on commons. It may also frequently be met with in tangled hedge-bottoms, especially those where ditches are made beside them, and haunts the gorse coverts and thickets near the moors, and even the brambles and briars upon their wide expanse. The bird is a most skulking little creature, rarely seen, but invariably detected by the thrilling chirping song, like that of the grasshopper, only louder and more prolonged. The birds pair soon after their arrival. Like other marsh Warblers the present species may be found in some numbers within a small area of suitable ground, but no social tendencies are evinced, especially after nest-building commences. The nest is either built upon the ground or only a short distance above it. It is frequently placed under brushwood, amongst tall, rank grass, especially in a tangled hedge-bottom, or on the ground in a tuft of grass in a plantation. Sometimes it is placed under a gorse bush. The nest is somewhat compact and deep, made externally of dry grass, moss, and dead leaves, and lined with finer round dry grass-stalks. If the first nest is taken, another, and even a third will be made, this accounting for the late nests of this species which some writers have interpreted as proof of a second brood. The sitting bird glides very quietly off the nest, and threading its way through the herbage and branches is seldom seen, except for a fleeting moment.

RANGE OF EGG COLOURATION AND MEASUREMENT: The eggs of the Grasshopper Warbler are from four to six or even seven in number. They are white, with a pinkish tinge in ground colour, profusely spotted and sprinkled with reddish-brown, and with similar underlying markings of violet-gray. The markings are usually most numerous on the larger end of the egg, and more bold and decided in character. Occasionally a few hair-like streaks of dark brown occur. In some eggs the markings are dark in tint, and are mostly distributed in an irregular zone; others are uniform pale brown. Average measurement, ·7 inch in length, by ·53 inch in breadth. Incubation, performed mostly by the female, lasts fourteen or fifteen days.

DIAGNOSTIC CHARACTERS: The general pink appearance and finely-dusted markings seem to distinguish the eggs of the Grasshopper Warbler from those of all other species breeding in the British Islands.

Although formerly a rare but regular summer migrant to the British Islands, Savi's Warbler (*Locustella luscinioides*) is now only too probably extinct as a breeding species. The extensive drainage of the fens and marshes of East Anglia has robbed the birds of their summer quarters, and compelled them and their descendants to seek haunts elsewhere. It formerly bred in Norfolk, Cambridgeshire, and Huntingdonshire, and possibly in one or two adjoining counties, arriving in April. The nest is carefully concealed amongst sedge, rushes or reeds, and made at varying heights from six inches to a yard above the sodden ground. It is a deep, cup-shaped structure almost entirely made of the flat ribbon-like leaves of the sedges, generally *Glyceria*, the narrow ones being used for the lining. The eggs (laid in May and June) are from four to six in number, and vary in ground colour from white or pale gray to pale buff, sprinkled and freckled over the entire surface with ashy-brown, and with underlying markings of violet-gray. The spots are most numerous round the larger end of the egg, where they often form a distinct zone. Sometimes a few hair-like streaks of darker brown occur. Average measurement, ·78 inch in length, by ·57 inch in breadth. They most nearly resemble those of the Grasshopper Warbler, but may generally be distinguished by their browner appearance. Both birds assist in the duties of incubation, which lasts about fourteen days. It is not known to rear more than one brood in the season.

Family TURDIDÆ. Genus TURDUS.
Sub-family *TURDINÆ.*

SONG THRUSH.

TURDUS MUSICUS, *Linnæus.*

Double Brooded. Breeding season, February to August.

BRITISH BREEDING AREA: The Song Thrush is generally distributed throughout the British Islands, breeding in all parts sufficiently wooded to meet its requirements. It breeds on many of the Hebrides, even on the Orkneys, but not on the Shetlands.

BREEDING HABITS: The Song Thrush is a more or less migratory species in most parts of our area, leaving its breeding-grounds in the late autumn, and returning very early the following season. The Song Thrush may be found breeding almost everywhere, especially in well-cultivated districts, in orchards, shrubberies, hedgerows, woods, coppices, and plantations. The bird pairs very early in the spring, and in open seasons the eggs are often laid by the end of February. The Song Thrush is not at all social during the nesting season, and although several nests may be found in a small area, each pair keeps to itself. The nest is placed in a great variety of situations, the earliest of the season by preference being built amongst evergreens. It is usually made well in the centre of a bush or hedge, often on the ground, at the root of the hedge-bushes, or on the top of a bank. Less frequently it is placed on a stone jutting from a wall covered with ivy, or amongst ivy on a tree-trunk; whilst evergreen shrubs, such as hollies, yews, myrtles, and laurels, are favourite sites. It is placed at heights varying from a few feet to as many as twenty feet or even more from the ground.

The nest is large, and made externally of dry grass, fine twigs, and sometimes a little moss or a few dead leaves. Then a lining of mud is inserted, which is finally finished off with a second layer of wet, rotten wood, obtained from decaying timber in the vicinity. Very often the nest remains empty for a day or so after it is finished. The Song Thrush is a close sitter, remaining brooding over the eggs until driven from them. When disturbed both parents often become very noisy and anxious for the safety of their home.

RANGE OF EGG COLOURATION AND MEASUREMENT: The eggs of the Song Thrush are four or five in number, very exceptionally six. They are of a clear turquoise-blue in ground colour, more or less thickly spotted and speckled with dark blackish-brown, and with a few underlying markings of similar character of ink-gray. The spots vary a good deal in size, some being round, others irregular in shape, and most are distributed over the larger end of the egg. Some varieties are blotched rather than spotted with paler brown; others are without markings at all. They also vary considerably in shape and size. Average measurement, 1·0 inch in length, by ·8 inch in breadth. Incubation, performed mostly by the female, lasts from thirteen to fifteen days.

DIAGNOSTIC CHARACTERS: The turquoise-blue ground colour and nearly black spots readily distinguish the eggs of the Song Thrush from those of all other British species. The nest also is unique.

Family TURDIDÆ.   Genus TURDUS.
Sub-family *TURDINÆ*.

MISSEL-THRUSH.

TURDUS VISCIVORUS, *Linnæus*.

Double Brooded. Breeding season, February to July.

BRITISH BREEDING AREA: The Missel-Thrush is widely distributed throughout the wooded districts of the British Islands, breeding as far north as the Orkneys, and in many of the Hebrides. A hundred years ago the Missel-Thrush appears to have been unknown in Ireland, but it is now widely dispersed and increasing. This species has also considerably extended its area in Scotland, following the planting of trees. I am of opinion that this Thrush breeds more abundantly in the northern counties of England than in the southern and south-western counties.

BREEDING HABITS: A marked migration of this species takes place in many of the northern and more exposed districts. The haunts of this fine Thrush are principally in well-timbered districts, woods, plantations, coppices, orchards, parks, pleasure-grounds, and farm lands where trees are common. In moorland districts it frequents the alder- and birch-fringed streams, and the larch and fir woods. The Missel-Thrush pairs very early in the year, and frequents a locality for weeks before nest-building commences. The nest is usually made at some considerable elevation, and is never seen in such lowly sites as that of the Song Thrush or the Blackbird. Sometimes several nests may be found within a very small area, but no social instincts are manifested during the breeding season. The nest is placed in a fork of the branches, often on a branch close to the stem, and rarely amongst the more slender twigs.

I have, however, known it made at the end of a long, slender yew branch. It is composed of twigs, coarse dry grass, chickweed, and often large masses of wool, cemented and lined, first with mud, then with a thick bed of very fine grass, much of it often green. Sometimes it is made externally of bog moss, with a few twigs intermingled, sometimes with green leaves studded here and there; but in all cases a lining of fine grass is inserted. Very often the nest of this Thrush is made in the most exposed and frequented situations, sometimes with a great tuft of wool hanging from the outside. The birds sit closely, but when disturbed are remarkably noisy and demonstrative, flying from branch to branch full of anxiety.

RANGE OF EGG COLOURATION AND MEASUREMENT: The eggs of the Missel-Thrush are four in number, never more, and rarely less. They vary from bluish or greenish-white to pale reddish-brown in ground colour, spotted, blotched, and freckled with rich purplish-brown, and with underlying markings of gray. The markings vary considerably in number and distribution, but are usually most frequent on the large end of the egg, where they frequently form a zone. If blotched the colour is paler. Average measurement, 1·25 inch in length, by ·86 inch in breadth. Incubation, performed by both sexes, lasts fifteen or sixteen days.

DIAGNOSTIC CHARACTERS: The size, buff or greenish ground colour, and purplish-brown spots combined, make the eggs of the Missel-Thrush easily identified.

Family TURDIDAE.   Genus MERULA.
Sub-family TURDINAE.

BLACKBIRD.

MERULA MERULA (*Linnæus*).

Double Brooded.   Breeding season, March to August.

BRITISH BREEDING AREA: The Blackbird is another widely distributed species, breeding commonly in most parts of the British Islands, even extending to some of the wildest upland districts, provided with cover in the form of orchards, gardens, and plantations. It breeds on some of the Hebrides and the Orkneys, and even on such barren spots as Ailsa Craig and the Bass. This species has largely extended its range of late years, following the planting of trees and cultivation generally.

BREEDING HABITS: For the most part the Blackbird is a resident in our islands, but a certain amount of migration takes place in some localities. The Blackbird may be met with breeding almost anywhere, provided some cover is to be found. Its favourite haunts are similar to those of the Song Thrush, and it is most abundant in well-cultivated districts, especially near houses, and in shrubberies, plantations, hedgerows, and orchards. It pairs early in the season, and is not at all social during the nesting period, although numbers of nests may be found within a comparatively small area. The nest is built in a variety of situations, and at different heights, some being on the ground, others but a few feet from it, and less frequently as many as forty feet. It is usually made well in the centre of a thick bush, often an evergreen, a yew, or a holly by preference, in a hedgerow, frequently on the bank, or amongst the roots of an old tree-stump, or in ivy, either on walls or timber. I have found it in a shed, and simply placed on a stone projecting from a bare wall. It is made of dry grass, sometimes a few twigs, and a little moss, and lined first with a thick plaster of mud or clay, and finally with dry grass. It is bulky, but somewhat shallow and very compact. The Blackbird is a close sitter, but is neither so noisy nor so demonstrative at the nest as the Song Thrush, the Missel-Thrush, or the Ring Ouzel.

RANGE OF EGG COLOURATION AND MEASUREMENT: The eggs of the Blackbird are from four to six, and, in very exceptional cases, eight in number. They vary in ground colour from pale blue to bright bluish-green, mottled, blotched, and spotted with reddish-brown, and with underlying markings of gray. On some varieties the markings are very handsome—bold blotches and splashes intermingled with smaller spots; on others they are closely mottled and freckled over the entire surface; on others they form a zone or circular patch on the larger end. Occasionally a few dark specks or streaks occur. A rarer variety is almost spotless and pale blue, or marked with a few lilac or pale brown spots and dashes. These are certainly not the produce of a union between a Blackbird and a Thrush, as has been suggested. The eggs of the Blackbird vary considerably in size and shape, some being very elongated, others almost globular. Average measurement, 1·2 inch in length, by ·85 inch in breadth. Incubation, performed by both sexes, lasts from thirteen to fifteen days.

DIAGNOSTIC CHARACTERS: It is impossible to give any character by which the eggs of the Blackbird can be distinguished from those of the Ring Ouzel. As a rule the breeding-grounds of the two species are different, and the noisy behaviour of the Ring Ouzel at the nest is very noteworthy. The Blackbird may occasionally nest on the breeding-grounds of the Ring Ouzel, but the Ring Ouzel never invades the normal area of the Blackbird, being confined strictly to moorlands and their immediate vicinity. With this exception the eggs of the present species cannot readily be confused with those of any other Thrush breeding within our area.

Family TURDIDÆ.
Sub-family TURDINÆ.

Genus MERULA.

RING OUZEL.

MERULA TORQUATA (*Linnæus*).

Single Brooded. Breeding season, April and May.

BRITISH BREEDING AREA: To a very great extent the range of the Ring Ouzel is similar to that of the Red Grouse, but it extends a little more to the south. It may be met with breeding on the Cornish Heights, on the moors of Devonshire and the wild hills of Somerset, the Cambrian Mountains, the Pennine Chain, the Cheviots, and northwards locally throughout Scotland, in districts suited to its requirements, including some of the Inner Hebrides, but not reaching either the Orkneys or the Shetlands. It is more sparingly and locally distributed throughout the mountain moorlands of Ireland. Instances of this species breeding in Kent, Suffolk, Norfolk, Leicestershire, and Warwickshire have been recorded, but such cases are very exceptional.

BREEDING HABITS: The Ring Ouzel is a summer migrant to our islands, reaching them in small numbers in March, but more generally in April. Its haunts are the wild moorlands, even up to the wind-swept summits of some of the higher peaks and tors. It may be found breeding on the rock-strewn hillsides studded with birches and gorse coverts and patches of bracken, or by the tumbling trout-streams that race down the slopes which lead up to the more level plateaux where the true moors begin. It arrives in flocks of varying size, but these soon disperse over the surrounding country in pairs, and nest-building commences almost directly afterwards. The nest is either made on the ground amongst the heath and ling, or in a low bush on a rocky bank. Frequently it is placed on the edge of a sloping bank, where a rough road has been cut by the moorland farmer for his sheep, or the keeper for his game-cart in August. The nest is very similar to that of the Blackbird, and is made on precisely the same plan. First a deep nest of dry grass and a few dead leaves is formed, often strengthened by some twigs; this is then well lined with a thick coating of mud; and finally a second lining of dry grass is added, the whole forming a very strong and compact structure. The birds are neither gregarious nor social during the nesting season, but several pairs may be found breeding within a small area, each keeping exclusively to themselves. When disturbed from the nest, even if it does not yet contain eggs, the parents become remarkably bold and noisy, flying about uttering their loud *tac-tac-tac*, and frequently sweeping past the observer's head.

RANGE OF EGG COLOURATION AND MEASUREMENT: The eggs of the Ring Ouzel are four or five in number, and so closely resemble those of the Blackbird in size, shape, and variation of colour, that it is quite needless to repeat their characteristics. Incubation, performed chiefly by the female, lasts fourteen or fifteen days.

DIAGNOSTIC CHARACTERS: **The eggs of this Ouzel**

cannot be distinguished from those of the Blackbird. They are best identified at the nest, where the actions of the parents make this matter an easy task. The Ring Ouzel is easily recognized by its conspicuous white gorget.

Family TURDIDÆ.   Genus ERITHACUS.
Sub-family *TURDINÆ*.

ROBIN.

ERITHACUS RUBECULA (*Linnæus*).

Double Brooded.   Breeding season, March to July.

BRITISH BREEDING AREA: Provided sufficient cover is offered, the Robin may be met with breeding more or less commonly throughout the British Islands, most abundant in well-cultivated districts and most local in upland areas, where its presence depends principally on whether human habitations surrounded with sufficient cover are to be found. It breeds as far north as the Orkneys, but has not yet reached the Shetlands as a nesting species. Like many other birds the Robin has increased its northern range within recent time, following the spread of tree-planting and general cultivation and *improvement*.

BREEDING HABITS: The Robin may be met with during the breeding season almost everywhere. It is to a certain extent a resibent, but some migration movement is noticeable, especially in the wilder and more northern districts. It may be found nesting in woods, plantations, shrubberies, coppices, lanes, gardens, orchards, and hedgerows; by the riverside, close to the sea-shore, amongst ruins, near houses and in farmyards —every description of haunt is alike frequented by the ubiquitous Robin. It is neither gregarious nor social, and jealously guards its haunt from all invasion. The Robin may pair for life, if returning season after season to one spot to nest is any indication, but after the breeding season is over the bird is remarkably solitary in its habits. The nest is placed in a well-sheltered situation, such as in a hole in a wall, amongst the roots of trees, far under an over-hanging bank, and in ivy either on the ground or growing over walls. Curious sites are by no means uncommon, such as the interior of a flower-pot, or a water-can, or in a shed, etc., etc. The nest is a large structure, made externally of moss, dry grass, quantities of dead leaves (oak by preference), and coarse roots, most of the material being in the front, the cup being as far back as possible, and neatly lined with finer roots, horsehair, and very rarely a few feathers. The Robin is a close sitter, and when disturbed will often perch close by and keep up a low piping note, uttered at intervals of a few seconds. It is much attached to a site, and will continue to occupy it notwithstanding much disturbance.

RANGE OF EGG COLOURATION AND MEASUREMENT: The eggs of the Robin are from five to eight in number (I have taken clutches of the latter amount), but usually six are found. They are white in ground colour, spotted and freckled with brownish-red, and with underlying markings of gray. Very often most of the markings form a zone, or less frequently a circular patch at the larger end of the egg; another common variety is so thickly mottled as to hide most of the ground colour. Rarer varieties are spotted with darker brown, and streaked with nearly black; or pure and spotless white. Average measurement, ·8 inch in length, by ·6 inch in breadth. Incubation, performed by both sexes, lasts on an average thirteen days, but the eggs are often sat upon as soon as laid.

DIAGNOSTIC CHARACTERS: The eggs of the Robin are best distinguished by their size, white ground colour, and brownish-red markings. The nest also is very characteristic, both in position and materials. The eggs cannot readily be confused with those of any other species breeding in our area.

Family TURDIDÆ.   Genus ERITHACUS.
Sub-family *TURDINÆ*.

NIGHTINGALE.

ERITHACUS LUSCINIA (*Linnæus*).

Single Brooded.   Laying season, May.

BRITISH BREEDING AREA: The distribution of the Nightingale in our islands is remarkably restricted, and confined almost exclusively to the area of the Plains: the Plain of York, the Central Plain, the Eastern Plain, the valley of the Thames, and the southern counties as far west as Shropshire and Somerset. It is said to breed sparingly and locally in Breconshire and Glamorganshire. The Nightingale is unknown in Scotland and Ireland.

BREEDING HABITS: The Nightingale arrives in the southern portions of our islands about the middle of April, but is at least a fortnight later in its northern haunts. Its principal haunts are in small woods, plantations, marshy spinneys, the outskirts of hop gardens, shrubberies, and orchards. To a certain extent the

Nightingale is social for a short time after its arrival, but as soon as the birds have paired they betake themselves to their respective haunts, from which they jealously drive intruders. The nest of this species is almost invariably made on the ground; although in Spain, where the temperature is so much higher, it has been found in hedges some five feet from the ground. The old haunts, to a great extent, seem to be resorted to each season. The nest is usually placed amongst the rank herbage of the woods or plantations, amongst old, exposed roots, on a bank, or in a hedge-bottom. Less frequently it is made amongst ivy, a foot or so from the ground, or amongst a drifted heap of dead leaves. The nest somewhat closely resembles that of the Robin, being a large structure composed of dry grass or bits of the flat leaves of rushes, moss, and quantities of dead leaves, usually those of the oak, and lined with roots, fine grass, and a little horsehair, more rarely with a scrap or two of vegetable down intermixed. The Nightingale is a close sitter, and is not demonstrative at the nest, slipping quietly off into the nearest cover, and skulking until the disturbance has passed.

RANGE OF EGG COLOURATION AND MEASUREMENT: The eggs of the Nightingale are from four to six in number, usually five, and vary from dark olive-brown to bluish-green. There are two somewhat distinct types of the eggs of this species. The olive-brown type is bluish-green in ground colour, which is almost concealed by the surface-marks of reddish-brown; the bluish-green type is only sparingly marked with reddish-brown, and frequently has most of that colour congregated in a circular patch at one or the other end of the egg. Rarely this patch of colour occurs at both ends. A few dark lines are occasionally seen on the eggs. The two types are produced by the greater or less amount of surface colouring. Average measurement, ·83 inch in length, by ·6 inch in breadth. Incubation, performed chiefly by the female, lasts fourteen days.

DIAGNOSTIC CHARACTERS : The eggs of the Nightingale cannot readily be confused with those of any other species breeding in our islands, their brown or olive appearance making their identification an easy task.

Family TURDIDÆ.
Sub-family *TURDINÆ.*

Genus RUTICILLA.

REDSTART.

RUTICILLA PHŒNICURUS (*Linnæus*).

Single Brooded. Laying season, May.

BRITISH BREEDING AREA : The Redstart breeds more or less locally in most parts of England, especially so in the south-west peninsula, from Devon to Cornwall. In Devon it is most abundant on migration. Everywhere it becomes rarer in the western counties, and is said only to penetrate in any number into Wales as far as Breconshire. In Scotland it becomes even more local and less plentiful, although it has been met with breeding as far north as the mainland extends. Its occurrence in Ireland is only abnormal. Its northern range appears to be increasing.

BREEDING HABITS : The Redstart arrives in our islands early in April, a week or so later in the more northern localities. The haunts of the Redstart are principally in or near cultivated districts. It frequents the borders of, and the open spaces in woodlands, coppices where plenty of hollow timber is to be found orchards, gardens, and the vicinity of ruins. I am of opinion that this species, in most cases at any rate, pairs for life, although the sexes do not migrate in company, the males arriving a few days before the females. Regularly every season the same nesting-place will be used, and that in spite of continual robbery and disturbance. This species is not at all social, each pair confining itself to a certain haunt. The nest is invariably made in a hole of some kind, either in a tree or wall, the side of a building, or in the crevice of a rock. Exceptionally a flower-pot, a pump, or other eccentric site is chosen. The hole is never altered in any way, and varies in depth from a few inches to a foot or more, and from a few feet to as many as twelve from the ground. The nest is a loosely-made structure of dry grass, moss, sometimes a little wool, and a few leaves, and lined with hair and plenty of feathers. The parent bird sits very closely, usually allowing itself to be lifted from the nest, but shows little anxiety and keeps away until the disturbance has passed. The birds become much more anxious when the young are hatched.

RANGE OF EGG COLOURATION AND MEASUREMENT: The eggs of the Redstart are from five to eight in number, six being an average clutch. They are uniform pale blue and somewhat finely polished. It is said that the eggs of this species are "occasionally speckled with reddish," but surely this must be a mistake! Average measurement, ·75 inch in length, by ·55 inch in breadth. Incubation, performed by both sexes, lasts fourteen days. The eggs of this species may be removed again and again, and they will be replaced by others. I have taken a dozen from one nest in a single season.

DIAGNOSTIC CHARACTERS: The uniform pale blue colour, fragile shell, and high polish, readily distinguish the eggs of the Redstart from those of the Hedge

Sparrow, the only ones with which they can be confused in our islands. The situation of the nest is also an unfailing guide to their correct identification.

The Black Redstart (*Ruticilla tithys*) has been said to breed in the British Islands, but there is, as yet, no positive evidence in support of the fact. This species is a fairly numerous winter visitor to our area, and it is not improbable that a few odd pairs may remain over the summer and breed with us. The eggs are usually laid in May. The nest is generally made in an outbuilding, in holes in walls, and in rock crevices. It is rather a bulky structure, loosely put together externally, but the cup containing the eggs is neat and remarkably well finished. The outer materials are composed of roots, dry grass, moss, straws, and dead stalks of plants; the cup lined with finer grass, moss, hair, and feathers. The eggs are from four to seven in number, five being an average clutch. They are normally pure white, but occasionally a brownish or bluish tinge is perceptible. Very exceptionally they are minutely and indistinctly speckled with brown on the larger end—a reversion probably to some ancestral colouration. Average measurement, ·75 inch in length, by ·58 inch in breadth. Incubation is performed by both sexes, but mostly by the female, and lasts about fourteen days. Two broods are generally reared in the year.

Family TURDIDÆ.　　　　　　　　　　　Genus SAXICOLA.
Sub-family TURDINÆ.

WHEATEAR.

SAXICOLA ŒNANTHE (*Linnæus*).

Single Brooded.　Laying season, April and May.

BRITISH BREEDING AREA: The Wheatear is widely but to some extent locally distributed throughout the British Islands, especially so in the southern and western counties of England. It becomes very perceptibly commoner northwards, and breeds in the Hebrides (even reaching St. Kilda), the Orkneys, and the Shetlands. In Ireland it is equally widely dispersed, but is more abundant in wild upland localities than in the more cultivated areas.

BREEDING HABITS: The Wheatear is one of our very earliest spring migrants, beginning to arrive towards the close of March, and continuing to do so throughout the following month. Its favourite summer haunts are open lands, stone quarries, warrens, moorlands, and mountain sides and downs. Like all the *stone* Chats the Wheatear shuns the timbered districts, and is rarely seen near trees or brushwood except during passage. Although gregarious during migration, the Wheatear is not even social during the nesting season, each pair keeping to its own particular haunt, notwithstanding the fact that many individuals may be breeding in a small area. This species pairs soon after the arrival of the females, the males preceding them by a few days. The nest of the Wheatear is invariably placed in a well-sheltered situation, and hence is very difficult to find. Sometimes it is made in a crevice of a rock, under a heap of stones, in a hole in a wall, or under a piece of loose turf in the rough fallows. Less frequently a deserted rabbit-burrow is selected; whilst in Scotland I have repeatedly found it in the side of a peat-stack. At St. Kilda a hole in the wall of the rough "cleats" or shelters for the sheep is selected. It is a loosely-made structure, composed externally of dry grass, roots, and sometimes a little moss, and lined with hair, fur, wool, finer roots, and feathers, according to which material is readiest to hand. The bird sits closely, and though restless and anxious enough when its breeding-haunt is invaded, rarely betrays the exact whereabouts of the nest.

RANGE OF EGG COLOURATION AND MEASUREMENT: The eggs of the Wheatear are from four to seven in number, five or six being most usually found. They are somewhat elongated in shape, very pale blue in ground colour, and generally without markings of any kind. Varieties, however, may be met with on which a few dust-like specks of purplish-brown occur, usually on the large end of the egg. Average measurement, ·82 inch in length, by ·62 inch in breadth. Incubation, performed chiefly by the female, lasts fourteen days. In spite of very positive statements to the contrary, I do not think the Wheatear ever rears more than one brood in the season.

DIAGNOSTIC CHARACTERS: The size and pale blue colour readily distinguish the eggs of the Wheatear from those of any other species likely to be confused with them in our islands. The nest also assists to their correct identification.

Family TURDIDÆ. Genus PRATINCOLA.
Sub-family TURDINÆ.

WHINCHAT.

PRATINCOLA RUBETRA (*Linnæus*).

Single Brooded. Laying season, May and June.

BRITISH BREEDING AREA: The Whinchat is fairly well distributed throughout Great Britain, but becomes rarer and more local in Scotland (although it certainly breeds as far north as Caithness and on some of the Hebrides) and in the south-west peninsula of England. In Ireland it is even more local and scarce, but its breeding area is by no means accurately defined.

BREEDING HABITS: The Whinchat arrives in the southern portions of our islands during the latter half of April, but is nearly a fortnight later in more northern districts. The principal breeding-haunts of this species are hay-meadows, commons, gorse coverts, and moorlands. A week or so after its arrival the Whinchat may be observed in pairs. It is not in any way a social species during the breeding season, and each pair keeps exclusively to one particular haunt until the young can fly. The nest is built in a variety of situations, according to the accommodation at hand. In the hay-meadows it is almost invariably placed on the ground, amongst the tall mowing-grass, at varying distances from the hedge-sides; elsewhere it is frequently at the foot of a gorse bush, or amongst heather or tall, rank herbage. A small cavity is scraped out when the nest is made on the ground. The home of the Whinchat is loosely put together, composed of dry grass and a little moss, and lined with fine roots and horsehair. The parent birds are excessively wary at the nest, flitting about from stem to stem, and persistently uttering their monotonous call-note (*ü-tac*), yet seldom if ever betraying its whereabouts. The best way to find the nest of this and other ground-building species is to walk across their haunts at night—a time when they sit close until almost trodden upon.

RANGE OF EGG COLOURATION AND MEASUREMENT: The eggs of the Whinchat are from four to six in number. They are a clear turquoise-blue in ground colour, faintly dusted with light brown, the markings generally forming a zone. In shape the eggs are somewhat peculiar, being almost as much pointed at one end as the other. Average measurement, ·75 inch in length, by ·58 inch in breadth. Incubation, performed chiefly by the female, lasts about fourteen days.

DIAGNOSTIC CHARACTERS: The elliptical shape, dark blue colour, and fine, almost invisible markings, readily distinguish the eggs of the Whinchat from those of the Stonechat, the only species with which they are at all likely to be confused in our islands. It might be remarked that these two species rarely inhabit the same districts.

Family TURDIDÆ. Genus PRATINCOLA.
Sub-family TURDINÆ.

STONECHAT.

PRATINCOLA RUBICOLA (*Linnæus*).

Single Brooded.   Laying season, April and May.

BRITISH BREEDING AREA: With the exception perhaps of the Orkneys and the Shetlands, St. Kilda, and a few other of the most remote and barren Hebrides, the Stonechat is pretty generally dispersed throughout the British Islands, although somewhat local and fastidious in the choice of a haunt. It is perhaps nowhere so common as the Whinchat, but certainly breeds in England more to the south-west than that species.

BREEDING HABITS: The Stonechat, although it may be subject to a considerable amount of internal migration, is a resident in the British Islands. Its principal breeding-haunts are furze coverts, heaths, commons, and rough, open, rock-strewn ground, studded with low bushes and thickets of briar and bramble, especially near to or adjoining the moors. In South Devon this species habitually frequents osier beds, its young being reared before the willows are much grown. It is probable that the Stonechat pairs for life, although the birds are more often than not met with solitary after the young are reared. The nest is either on the ground or at most a few inches above it, according to the site selected. It is frequently placed at the foot of a gorse bush, amongst heath and other rank herbage, or amongst the long, tangled grass growing round the stumps of the osiers, occasionally on the stump itself, hidden by brambles and other interlaced vegetation. The nest is extremely difficult to find, unless stumbled upon by accident. It is a simple little structure, loosely fabricated, made externally of dry grass, moss, and roots, and lined with finer grass, horsehair, feathers, and less frequently wool. The birds are very wary at the nesting-place, and although they show themselves continually and flit tirelessly from spray to spray, occasionally indulging in longer flights, and repeatedly uttering their double call-note (*ü-tic*), they are most careful not to betray their secret.

RANGE OF EGG COLOURATION AND MEASUREMENT: The eggs of the Stonechat are from four to six in number. They are pale bluish-green in ground colour, spotted, blotched, and freckled with reddish-brown. The markings generally form a zone round the larger end of the egg, and in some varieties almost entirely cover the end. Rare varieties are almost without any markings; more frequently a richly-marked type is seen, where the spots are large and blotchy. Average measurement, ·7 inch in length, by ·58 inch in breadth. Incubation, performed chiefly by the female, lasts about fourteen days.

DIAGNOSTIC CHARACTERS: The paler ground colour and much richer and more clearly defined markings distinguish the eggs of this species from those of the Whinchat. The unspotted varieties require more care in identification, but the locality of the nest is then of some importance.

Family TURDIDÆ.   Genus ACCENTOR.
Sub-family ACCENTORINÆ.

HEDGE ACCENTOR.

ACCENTOR MODULARIS (*Linnæus*).

Double Brooded.   Laying season, February to July.

BRITISH BREEDING AREA: With the exception of a few of the most barren of the Hebrides, and the Orkneys and the Shetlands, the Hedge Accentor is very generally distributed throughout the British Islands, breeding in every county in more or less abundance, becoming most plentiful, however, in cultivated districts. It has considerably increased its range northwards with the progress of cultivation and tree-planting.

BREEDING HABITS: Like the ubiquitous Robin the Hedge Accentor may be met with breeding almost everywhere, especially in sheltered, well-timbered, and cultivated districts. It is a resident in our islands, and does not wander far from its usual haunts even in winter, save in the most bleak and unsheltered areas. Its favourite breeding-grounds are hedgerows, shrubberies, gardens, and low underwood of all kinds. The Hedge Accentor pairs very early in the year, many doing so in January, and months before nest-building commences. It is one of the very earliest species to breed, and is unsocial and solitary right through the period. Several nests may frequently be found in a single hedge, yet each pair keeps exclusively to itself. The nest is placed in a great variety of situations, though never very high from the ground. Hedgerows are specially favoured places, as are also low thickets and dense bushes, ivy, either growing on timber, walls, or palings; heaps of hedge-clippings, briars, and brambles. Few nests are prettier than the charming little rustic cradle of the Hedge Accentor. Externally it is made chiefly of moss (though sometimes dry grass predominates), a few dead leaves and a little dry grass, amongst which a varying number of twigs are interwoven ; it is warmly lined with hair, wool, and feathers. I have a nest made externally chiefly of withered leaves and a few bits of lichen, bound together with round, dry grass-stalks and two or three dead twigs ; it is lined principally with moss and a little hair. The Hedge Accentor sits very closely, and when driven off glides very quietly into the cover, without demonstration of any kind. Nests of this species are often made in very exposed and frequented situations.

RANGE OF EGG COLOURATION AND MEASUREMENT: The eggs of the Hedge Accentor are from four to six in number. They are a clear dark turquoise-blue, and without markings. Average measurement, ·77 inch in length, by ·6 inch in breadth. Incubation, performed by both sexes, lasts from twelve to fourteen days.

DIAGNOSTIC CHARACTERS: The intense blue colour and absence of spots, combined with the open-sited nest, distinguish the eggs of this species from those of all others likely to be confused with them.

Family CINCLIDÆ. Genus CINCLUS.

DIPPER.

CINCLUS AQUATICUS, *Bechstein.*

Double Brooded.   Breeding season, April to July.

BRITISH BREEDING AREA: The breeding range of the Dipper very closely follows that of the Ring Ouzel, extending from Cornwall, Devon, and Somerset northwards through Wales and the Pennine Chain to the Border. In Scotland the bird is much more widely dispersed, owing to the more favourable nature of the country, and breeds throughout that area, extending to the Outer Hebrides and the Orkneys. In Ireland it is equally widely dispersed throughout all the mountainous districts suited to its requirements.

BREEDING HABITS: The Dipper is a resident in our islands, although there is an appreciable amount of internal migration, subject to fluctuations of weather. The haunts of the Dipper are the wild, rapid-flowing mountain streams and brooks, where the water splashes over big boulders and the banks are rugged. Trees often fringe the banks of such torrents, but they are not essential to the Dipper's presence. The bird is neither a gregarious nor a social one, each pair appearing to have a vested right in a certain length of the stream, to which they keep pretty closely. It is most probable that the Dipper pairs for life, and regularly resorts each season to a favourite nesting-place. The nest is seldom or never at any distance from the water, and often so close to the stream as to be washed by the spray. It is usually made in a crevice of the rocks, on the side of a gorge, under a bridge, or in the masonry of a sluice or weir, or amongst the exposed roots of a tree; occasionally it is even made in a crevice of the rocks behind a waterfall. The nest is a beautiful structure, globular, similar to that of the Wren, with a hole in the side for entrance. Externally the large nest (sometimes eighteen inches in length) is made almost exclusively of green moss, which often harmonizes closely with surrounding tints, strengthened here and there, especially round the hole, with dry grass; inside another nest, of dry grass, roots, and twigs, and lined with quantities of withered leaves arranged layer upon layer, is formed. The Dipper sits closely, often allowing itself to be taken from the nest, and when flushed from it makes little or no demonstration.

RANGE OF EGG COLOURATION AND MEASUREMENT: The eggs of the Dipper are from four to six in number, and are pure white and spotless. Average measurement, 1·0 inch in length, by ·75 inch in breadth. Incubation, performed by both sexes, lasts fourteen or fifteen days.

DIAGNOSTIC CHARACTERS: The eggs of the Dipper are best distinguished from those of the Kingfisher and the Great Spotted Woodpecker by their want of gloss, the shell being somewhat rough in texture. The situation of the nest, combined with the colour and size of the eggs, prevent any possible confusion.

Family TROGLODYTIDÆ.   Genus TROGLODYTES.

COMMON WREN.

TROGLODYTES PARVULUS, *Koch.*

Double Brooded.   Laying season, April to June.

BRITISH BREEDING AREA: The Common Wren is very widely and generally distributed throughout the British Islands, even extending to the Outer Hebrides (but not to St. Kilda), to the Orkneys and the Shetlands. It is of course much more abundant in the lowland, well-cultivated districts than in the more upland, treeless wilds.

BREEDING HABITS: The Wren is everywhere a resident in our islands. Its haunts are varied in the extreme, from the farmyard and the garden round the homestead to the bare, treeless wilderness of the crofter's cottage or the Irish bothy. It frequents woods, plantations, coppices, spinneys, and shrubberies just as much as thickets, the brushwood on the banks of streams, sunk fences, ditches, hedgerows, and the rough broken ground clothed with gorse and bramble and fern near the moors, and even the long ling and heath on those wide wastes. The Wren pairs annually in some cases, and rather early in spring, although it is not improbable that many birds are united for life, and yearly breed in one spot. It is not at all a social species, each pair keeping to a particular haunt, and as soon as the breeding season is over is most often met with solitary and alone. The nest is a charming little structure, globular in shape, with a small hole in the front, often near the top, and frequently, when in bushes, on the side. It is built in a great variety of situations—in thick bushes of all kinds, in brambles and amongst ivy, in the sides of haystacks, and frequently suspended from a long slender branch, usually of an evergreen, a yew by preference. It is also often made under an overhanging bank, wedged close in a crevice of bare soil or rock; amongst thatch, especially the roof of a linhay, in wood-stacks, and amongst exposed roots of trees, and in stumps in the hedgerows. As the Wren depends a good deal on mimicry for the concealment of its nest, the external materials vary a good deal in character. According to circumstances, the nest is made outwardly of moss, dead fronds of fern, withered leaves, dry grass, and lichens; internally it is composed of moss, hair, and feathers. The latter materials, however, are not always used, and vary a good deal in quantity. Round the entrance-hole a few straws, round dry grass-stalks, roots, and twigs are interwoven with the other materials, thus strengthening the structure in the part which is subject to the most wear and tear. The female alone is the architect, and often a fortnight is taken to build the nest. The Wren is a close sitter, and often becomes very noisy when the nest is menaced. This species will forsake its unfinished nest more readily than any other bird known to me.

RANGE OF EGG COLOURATION AND MEASUREMENT: The eggs of the Wren are from four to six, and more rarely eight, in number (sixteen young birds, however, have been recorded from a single nest, which seems almost incredulous). They are white in ground colour, more or less sparingly marked with brownish-red and grayish-brown. The markings vary a good deal in amount, some eggs are spotless, others have few and very faint markings, whilst more rarely varieties are handsomely freckled with spots and specks, most of them usually forming a zone round the larger end. Average measurement, ·7 inch in length, by ·52 inch in breadth. They vary considerably in size and shape. Incubation, performed chiefly by the female, lasts thirteen or fourteen days.

DIAGNOSTIC CHARACTERS: It is impossible to give any reliable characters that will enable the student to distinguish the eggs of the Wren from those of the Willow Wren, Great Titmouse, Nuthatch, and similar species. The form and character of the nest is the most unfailing guide to their correct identification.

Family TROGLODYTIDÆ.   Genus TROGLODYTES.

ST. KILDA WREN.

TROGLODYTES PARVULUS HIRTENSIS, *Seebohm.*

Probably Double Brooded. Laying season, April to June.

BRITISH BREEDING AREA: Notwithstanding Mr. Dresser's contention to the contrary, I stoutly maintain the sub-specific distinctness of the present bird from the Common Wren. Naturalists belonging to the older school of philosophy, burdened and biased as they are by all the præ-Darwinian ideas on local variation and sub-specific characters (which amount practically to a complete ignoring of them), cannot be expected to change their opinions, or to train their warped and defective powers of vision and perception to a correct focus of those facts which Nature unquestionably presents to us in bewildering plenitude; but one would have thought that the man who was a party to the separation of *Parus britannicus* from *Parus ater* more than twenty years ago, would by this time have had sufficient experience of sub-specific forms not quite so rashly or inconsistently to ignore them. *Troglodytes parvulus hirtensis* lives and, I hope, flourishes still, in spite of Mr. Dresser's no doubt well-intentioned efforts to pluck it from the tree of ornithological life, prompted as they were more by eagerness to inflict a thrust at its humble discoverer, than by any laudable desire for a correct statement of facts, or arrival at truth; especially when we remark his readiness to describe yet another full-blown species (not sub-species, mind you, such an expression has neither use nor meaning to the old school of ornithologists) of Coal Titmouse from Cyprus, under the very euphonious and classical name of *Parus cypriotes!* One would have wished that he had been equally consistent in his masterly treatment of the *Saxicolæ;* in his unwarrantable amalgamation of *Parus teneriffæ* with *Parus ultramarinus;* and similarly discerning in the case of the Marsh Titmouse, British examples of which, however dubiously, now go down to posterity (wrapped in an ill-concealed sarcasm), thanks to the acumen of Dr. Stejneger, under the name of *Parus palustris dresseri!*

But to return to the breeding area of the St. Kilda Wren. So far as is known this insular form of the Common Wren is confined to the group of Atlantic Islands known collectively as St. Kilda. And here I may remark that "the few pairs which inhabit the island" are by no means "extirpated" or even likely to be.

BREEDING HABITS: The St. Kilda Wren is a resident on the islands whose name it bears. It may be met with in every part of these bare treeless islets, where not a single bush of any kind flourishes. It hops about the rocks and on the face of the broken sea-cliffs; dodges about the rough walls that divide the crofters' gardens from the grassy downs, and frequents the "cleats" or sheep shelters. The bird very probably pairs for life, and makes its home in one chosen spot season after season, although building a new nest each successive year. The nests of this species that I have seen were made in crevices of the rough walls, in heaps of stones, and under the turf and stone roofs of the "cleats." The nest very closely resembles that of the Common Wren, being made on precisely the same model, globular, with a small hole at the front, near the top, or in the side. It is composed almost entirely of moss, but round the entrance-hole, especially just below it, a few dry grass-stalks are interwoven, and is lined profusely with feathers and a little horsehair, the latter being pulled from the numerous Puffin snares set in the cliffs. The birds were remarkably tame at the nest, going in and out as I stood watching them. The female sits closely, often allowing herself to be lifted from the eggs.

RANGE OF EGG COLOURATION AND MEASUREMENT: The eggs of the St. Kilda Wren are six in number. They are pure white in ground colour, profusely spotted, especially round the larger end, with brownish-red, and a few underlying markings of paler and grayer brown. Some varieties are almost devoid of markings; others have the markings in a circular mass at the end. Average measurement, ·72 inch in length, by ·57 inch in breadth. Incubation, performed chiefly by the female, lasts fourteen days.

DIAGNOSTIC CHARACTERS: It is impossible to give any character that will distinguish the eggs of this sub-species, either from those of its near ally (but on an average they are a little larger), or from those of the Great Titmouse, Nuthatch, etc. The locality is a safe guide to their correct identification and authentication, as no other species breeds on St. Kilda whose eggs can be confused with them.

Family MUSCICAPIDÆ.   Genus MUSCICAPA.

SPOTTED FLYCATCHER.

MUSCICAPA GRISOLA, *Linnæus*.

Single Brooded.   Laying season, end of May or June.

BRITISH BREEDING AREA: The Spotted Flycatcher is a wide-ranging species, and more or less commonly distributed throughout the British Islands, nesting in every part suited to its requirements. It does not appear to visit the Hebrides (Skye excepted), nor to breed in the Orkneys and Shetlands, and becomes rarer and more local in the extreme north.

BREEDING HABITS: The Spotted Flycatcher is a late migrant to our islands, arriving in its old haunts early in May. Its principal breeding-haunts are the outskirts of woods, well-timbered parks, gardens, pleasure-grounds (it still breeds in some of the London parks and open spaces), orchards, and groves of trees. It is most probable that this species pairs for life, and yearly returns to an old accustomed spot to breed. It is not in any way social, each pair keeping exclusively to themselves in one particular haunt. The nest is made in a great variety of situations, but generally in a snug nook. It is often built on the horizontal branch of a fruit tree trained along the wall, or on trellis-work supporting roses and creepers on the house side. My old friend, the late Mr. Milner of Meersbrook Hall, proudly pointed out to me a nest of this species made in a ledge over his front door, and he assured me that the birds built there every year. Other favourite sites are in crevices of the bark, especially in wych elms, or in shallow knot-holes, in holes in walls, or on a beam supporting the roof of a shed. Many curious sites have been known, such as the interior of a pump, on the hinge of a door, etc. The nest is small and somewhat loosely put together. Externally it is made of dry grass and moss, bound together with cobwebs and wings of various insects, and perhaps one or two rather large feathers; internally it is composed of roots, hair, and feathers. Considerable variation may be noticed in the nest of this species. Some nests contain twine, worsted, and cotton; others are garnished with lichens or dead leaves; some are entirely lined with roots, hair, or feathers respectively. The female sits closely, but is not very demonstrative when the nest is menaced.

RANGE OF EGG COLOURATION AND MEASUREMENT: The eggs of the Spotted Flycatcher are from four to six in number. They vary in ground colour from bluish-white to clear pea-green, blotched, freckled, and spotted with various shades of reddish-brown. Various well-marked types occur. On some the spots are large, semi-confluent, and nearly conceal the ground colour; on others they form a zone; a rarer and very beautiful variety is clouded all over with pale red. Average measurement, ·75 inch in length, by ·56 inch in breadth. Incubation, performed chiefly by the female, lasts about thirteen days, but very often the first egg is sat upon as soon as laid.

DIAGNOSTIC CHARACTERS: The bluish or greenish ground colour and the reddish-brown markings distinguish the eggs of the Spotted Flycatcher from those of other species. Some varieties, however, closely resemble certain types of those of the Robin. The situation and construction of the nest, however, prevent confusion.

Family MUSCICAPIDÆ.   Genus MUSCICAPA.

PIED FLYCATCHER.

MUSCICAPA ATRICAPILLA, *Linnæus*.

Single Brooded.   Laying season, May.

BRITISH BREEDING AREA: The Pied Flycatcher is another very local species, principally confined to the Cambrian Mountains, the Pennine Chain, and the Southern Highlands. It is known to breed regularly in Breconshire, Denbigh, and Merioneth, and the English counties on the Welsh border, thence in Lancashire, West Yorkshire, Durham, Westmoreland, Cumberland, and Northumberland across the Border, northwards to Caithness and Inverness-shire. Only breeds accidentally in such localities as Surrey, Middlesex, Oxfordshire, Gloucestershire, Hampshire, Dorset, Somerset, and Devonshire. Of only accidental occurrence in Ireland, not breeding.

BREEDING HABITS: The Pied Flycatcher arrives in our islands during the latter half of April. Its haunts are in the wilder and more romantic districts, although in other countries it is as much addicted to gardens and orchards as its commoner ally the Spotted Flycatcher. In our islands its breeding-grounds are principally situated in the more open woods and coppices, and the well-timbered banks of lakes and streams. It is probable that this species pairs for life, as in many cases the same nesting-places are resorted to yearly. No social tendencies are displayed, each pair keeping to themselves. The nest is invariably made in a covered site—a hole in a tree by preference, otherwise a hole or crevice in a wall or rock. I found a nest of this species in the

Rivelin valley, in South Yorkshire, in a rotten stem of a silver birch, the wood so brittle as to crumble away like paper. The hole is at various heights from the ground, sometimes but a few feet, at others as much as twenty feet. The nest is a loosely-made structure, composed of dry grass, dead leaves, and moss, lined either with fine roots, horsehair, wool, or feathers—sometimes with two or more of these substances. The birds never make a hole for themselves, neither do they alter a selected site in any way. The bird is a close sitter, often allowing itself to be taken from the eggs; it is not at all demonstrative at the nest.

RANGE OF EGG COLOURATION AND MEASUREMENT: The eggs of the Pied Flycatcher are from five to eight or even nine in number. They are uniform pale blue in colour. It is said that sometimes a few spots occur, but I have never seen anything of the kind. Average measurement, ·7 inch in length, by ·54 inch in breadth. Incubation, performed by both sexes, lasts fourteen days.

DIAGNOSTIC CHARACTERS: It is impossible to distinguish eggs of the Pied Flycatcher from those of the Redstart, so that they require the greatest care in their identification.

Family HIRUNDINIDÆ.   Genus HIRUNDO.

BARN SWALLOW.

HIRUNDO RUSTICA, *Linnæus*.

Double Brooded. Laying season, May and July.

BRITISH BREEDING AREA: With the exception of the Outer Hebrides, the Orkneys and the Shetlands (where it only exceptionally breeds), the Barn Swallow breeds throughout the British Islands in all localities suited to its requirements, becoming, however, more local and less numerous in the wilder and northern districts. It is not improbable that this species may breed occasionally in St. Kilda (conf. *Ibis*, 1885, p. 360).

BREEDING HABITS: The Swallow is a summer visitor to our islands, arriving from the second or third week in April onwards to the beginning of May, according to latitude. It is most abundant in well-cultivated districts, in country villages, near farmsteads and outbuildings, but is by no means rare even in wild uplands in the neighbourhood of the few houses that there occur. In my opinion the Swallow pairs for life, and most readers have probably remarked its annual return to certain favourite spots to breed ; not, however, using the same nest each season, as the House Martin generally does. The usual site for the nest of this Swallow is in or on a building of some kind, although there can be little doubt that holes in trees and rocks and caves were formerly its accustomed situation. Even now it occasionally resorts to these places, more frequently abroad, however, than in our densely-populated islands. The Swallow is more or less gregarious, and breeds in colonies of varying size, according to the amount of accommodation offered. In our islands the nest is usually made in a shed, barn, or other building, on a rafter supporting the roof, or on a stone projecting from the wall, or on a ledge close to the under surface of the tiles. A ledge in a chimney, or even in a much-frequented passage, is sometimes selected. The nest is composed chiefly of mud mixed with bits of straw, open, and lined with dry grass and feathers. In many Continental districts, however, it is made on the same model as that of the House Martin, only more open, and generally attached to a wall in the well-known manner. The Swallow is not a very close sitter, and betrays little anxiety when its nest is menaced, beyond occasionally flying in and out of the building.

RANGE OF EGG COLOURATION AND MEASUREMENT: The eggs of the Swallow are from four to six in number. They are pure white in ground colour, freckled, spotted, and blotched with various shades of rich coffee-brown, and with underlying markings of violet-gray. Usually the markings are pretty evenly distributed, but most thickly so on the larger end, often forming an irregular zone. Some eggs are handsomely blotched, others more finely speckled; some have the gray markings few and indistinct; others display them more numerous than the brown surface-spots. Average measurement, ·82 inch in length, by ·54 inch in breadth. Incubation, performed chiefly by the female, lasts from thirteen to fifteen days.

DIAGNOSTIC CHARACTERS: The rich coffee-brown and violet-gray spots on a white ground readily distinguish the eggs of the Swallow from those of all other species breeding in our islands. The situation of the nest is also very characteristic.

Family HIRUNDINIDÆ.   Genus CHELIDON.

HOUSE MARTIN.

CHELIDON URBICA (*Linnæus*).

Double Brooded.   Laying season, May and July.

BRITISH BREEDING AREA: The House Martin is widely distributed throughout the British Islands, breeding in almost every part, with the exception of the Outer Hebrides, but becoming a little less abundant and more local in various districts of Scotland, especially in the north, whilst in Ireland it is said not to be so common anywhere as the Swallow.

BREEDING HABITS: The House Martin is a summer migrant to our islands, reaching them in April, a few days later, as a rule, than the Swallow. The haunts of this charming little species are very similar to those of the Swallow, but it is even more ubique and cosmopolitan in its distribution, not only frequenting buildings of all kinds even in towns, but resorting to many ocean cliffs and inland precipices, far removed from the dwellings of man. There can be no doubt whatever that the House Martin pairs for life, and returns each season to its old nesting-place. It is also even more gregarious than the Swallow, some of its colonies being very large, certain buildings or cliffs being lined with nests. It is difficult to say which are the more favoured breeding-places of the House Martin—buildings or cliffs, for each seems almost as frequently chosen. When on buildings the nest is placed against the wall, under the gutter of the roof, in a corner of, or above windows, in fact anywhere where a projecting ledge affords some sort of shelter above it. On the cliffs the nests are made in various nooks and corners, sometimes in the roof of a large excavation where the rocks overhang; sometimes on the smooth, bare face, but invariably under a projection of some kind. The nest is made externally of mud, a semi-globular structure, shaped like half a basin, in some cases more elongated than others. This mud is brought in small pellets and built on piece by piece, sometimes with bits of straw intermixed, to give it better adhesive qualities. The narrow, but rather wide entrance-hole is at the top, either at the front or on one side. The inside of this mud shell is lined with dry grass and feathers. The birds are very much attached to their breeding-places, and will build nest after nest in one chosen spot, as soon as they are removed.

RANGE OF EGG COLOURATION AND MEASUREMENT: The eggs of the House Martin are four or five, rarely six in number. They are pure and spotless white, smooth in texture, and somewhat glossy. Average measurement, ·77 inch in length, by ·54 inch in breadth. Incubation, performed by both sexes, lasts thirteen days.

DIAGNOSTIC CHARACTERS: The larger size and more polished surface of the eggs of the House Martin distinguish them from those of the Sand Martin, the only species with which they are likely to be confused. The nests of the two species, however, are very different.

Family HIRUNDINIDÆ. Genus COTYLE.

SAND MARTIN.

COTYLE RIPARIA (*Linnæus*).

Double Brooded. Laying season, May and July.

BRITISH BREEDING AREA: The Sand Martin, if somewhat locally, is widely distributed throughout the British Islands, breeding in every part suited to its requirements, including the Outer Hebrides (but not St. Kilda), the Orkneys, and the Shetlands.

BREEDING HABITS: The Sand Martin is the earliest of the Swallows to arrive in our islands in spring, reaching them occasionally at the end of March, and more frequently during the first few days of April. The haunts of the Sand Martin principally depend on the presence of suitable cliffs and banks into which the birds can burrow and make their nests. Wherever such localities—soft earth cliffs inland or on the coast, sand-pits, quarries, railway-cuttings and the like—are to be found, the Sand Martin is usually present, although, be it remarked, it is somewhat capricious in its choice. The presence of water is preferred, but is not absolutely essential. The Sand Martin pairs for life, and yearly returns to breed in certain spots. It is a gregarious bird, some of its colonies being very extensive. A very noteworthy one is in a sand-cliff just outside Retford Station on the Great Northern Railway. The nest of the Sand Martin is placed at the end of a tunnel of varying depth, excavated by the bird in the soft bank or cliff. Both birds assist in making this gallery, which slopes upwards slightly from the entrance, and frequently turns several times before the end is reached. I have often known a big stone to stop the way, and cause the hole to be deserted. The holes vary in length from two to four feet, and are three or four inches in diameter. The end of the tunnel is widened out into a little chamber about six or eight inches high, and in this the nest is formed. It is a slight, loosely-fabricated structure of dry grass and straws, lined with a few feathers. A colony of Sand Martins is always a pretty sight, the birds gliding to and fro before their nest-holes, and entering and leaving them with little shyness.

RANGE OF EGG COLOURATION AND MEASUREMENT: The eggs of the Sand Martin are four or five, rarely six in number. They are pure white and spotless, but not quite so polished as those of the preceding species. Average measurement, ·72 inch in length by ·48 inch in breadth. Incubation, performed by both sexes, lasts thirteen or fourteen days.

DIAGNOSTIC CHARACTERS: The eggs of the Sand Martin are smaller and not so polished as those of the House Martin, the only species with which they are likely to be confused. The situation of the nest, however, prevents the slightest possibility of confusion.

Family PICIDÆ.   Genus IYNX.

WRYNECK.

IYNX TORQUILLA, *Linnæus*.

Single Brooded.   Laying season, May and early June.

BRITISH BREEDING AREA: The Wryneck is another local species, principally confined to the southern and eastern counties of England. It becomes rare in Wales

and the west of England generally, as well as north of
the Don valley in South Yorkshire. Its breeding range (if
any) in Scotland is imperfectly defined, whilst to Ireland
the bird is only a rare straggler on abnormal flight.

BREEDING HABITS : The Wryneck reaches our islands
in spring towards the end of March, or during the first
half of April. Its favourite haunts are parks, the more
open woodlands, well-timbered fields, orchards, gardens,
and coppices, in which plenty of decayed trees occur.
It is more attached to open timbered countries than to
thick woods. In the Rivelin valley this bird frequents,
or used to do so a dozen years ago, the rough, rock-strewn
slopes near Hollow Meadows, studded with clumps of
bushes and trees, on the very borders of the moors. I
am of opinion that the Wryneck pairs for life, and season
by season may be found breeding in one particular spot.
It is not at all sociable in its habits, and each pair keep
strictly to themselves. The shrill noisy cry of this bird
soon proclaims its presence in a district. Shortly after
arrival, the birds may be seen toying with each other
near the old nesting-place. This is almost invariably
in a hole in a tree, but very exceptionally it is said to
be in a bank. The bird does not excavate the hole
itself, although it may occasionally alter it slightly for
the purpose. A hole in any kind of tree will be selected,
often in a rotten stump, and varies considerably not only
in depth, but in height from the ground. The eggs are
laid on the powdered wood at the bottom of the hole,
no nest of any kind being provided for them. The
Wryneck sits closely, usually allowing itself to be taken
from the eggs, and will not only hiss when disturbed,
but sham death in a very extraordinary way.

RANGE OF EGG COLOURATION AND MEASUREMENT :
The eggs of the Wryneck are from six to ten in number.
They are pure white without markings, and are smooth

and polished. Average measurement, ·83 inch in length, by ·62 inch in breadth. Incubation, performed by both sexes, lasts fourteen days. The Wryneck will continue laying egg after egg in the same nest just as regularly as they are removed, as many as forty-two having been taken in each of two successive seasons.

DIAGNOSTIC CHARACTERS: The eggs of the Wryneck are most likely to be confused with large examples of the eggs of the Lesser Spotted Woodpecker, but they may be generally distinguished by their constantly larger size and lesser amount of polish. As the two species breed in very similar situations, great care should be exercised in their identification.

Family PICIDÆ. Genus GECINUS.

GREEN WOODPECKER.

GECINUS VIRIDIS (*Linnæus*).

Single Brooded. Laying season, April and May.

BRITISH BREEDING AREA: The Green Woodpecker is pretty generally distributed throughout the well-timbered districts of England, from Yorkshire southwards. It is rarer and more local in the north than in the south, and breeds very irregularly in the four northern counties. It becomes common in most of the counties adjoining the south coast, especially so in Devon and Cornwall. This Woodpecker is a very rare straggler to Scotland and Ireland, and does not breed in either of those countries.

Breeding Habits: The Green Woodpecker is a resident in our islands, frequenting those districts where timber is abundant, such as parks, old forest country, woods, and fields in which plenty of trees occur. It may, however, occasionally be met with in localities where trees are scarce, doubtless tempted by some favourite food. In South Devonshire it may be frequently seen on the sea-cliffs, apparently rarely going inland to the trees and woods. The Green Woodpecker pairs for life, although it is one of those species that the sexes do not keep very close company after the breeding season is over, being generally met with solitary. Neither is it at all social during the nesting period, each pair keeping to themselves. Sometimes the same nesting-place is used season by season; sometimes a new one is prepared, especially if the old abode has become water-logged, or been taken possession of by some other species. The Green Woodpecker nests in a hole in a tree, which the parent birds excavate for themselves, usually leaving a portion of the refuse of their work on the ground below. I have, however, known this bird nest in a hole in a cliff, between Paignton and Torquay—a large burrow, almost like a fox-earth, a few yards from the summit, but certainly not made by the Woodpeckers. The hole is circular, and very neatly made, either in a branch or in the trunk, where the wood is soft. In some cases the birds will bore through a thin layer of sound timber to reach the rotten portion. For a little distance the hole is horizontal, then a perpendicular shaft is dug out about a foot in depth, at the bottom of which it is enlarged into a kind of chamber, in which the eggs are deposited, on no other bed than is afforded by the dry dust and few wood chips. The bird is a close sitter, but makes little or no demonstration at the nest when disturbed.

RANGE OF EGG COLOURATION AND MEASUREMENT: The eggs of the Green Woodpecker are from five to eight in number, pure and glossy white, smooth and highly polished. Average measurement, 1·3 inch in length, by ·9 inch in breadth. Incubation, performed by both sexes, lasts, according to Thienemann, from sixteen to eighteen days.

DIAGNOSTIC CHARACTERS: The size, polished surface, and white colour, readily distinguish the eggs of the Green Woodpecker from those of all other species breeding in our islands, likely to be confused with them.

Family PICIDÆ.　　　　　　　　　　　　　Genus PICUS.

LESSER SPOTTED WOODPECKER.

PICUS MINOR, *Linnæus*.

Single Brooded, generally. Laying season, end of April and through May.

BRITISH BREEDING AREA: The range of the Lesser Spotted Woodpecker in our islands is very similar to that of the preceding species. The present bird is pretty generally distributed throughout the wooded districts of England, from Yorkshire southwards. It is most abundant in the southern districts, although everywhere somewhat local. It is not known to breed in Scotland or Ireland, only being a rare straggler to those countries.

BREEDING HABITS: The Lesser Spotted Woodpecker is a resident. It frequents very similar localities to those selected by the Green Woodpecker—woods, parks, pleasure-grounds, well-timbered fields and orchards. It is more often seen in the slender, topmost branches, however, than its larger ally, and from its shy, retiring disposition, is very often overlooked. It most probably pairs for life, although it is generally met with alone after the breeding season is over, and either returns to the same nesting-place yearly, or makes a new home near the old one. It is not at all social during the nesting period, each pair keeping to one particular haunt. The nest is in a hole in a tree, either of a branch or the trunk, at various heights from the ground. Frequently a pollard willow is selected, sometimes a dead stump in the tall hedges, or a fruit tree in the orchard; at other times a hole is made in the branch of a tall beech or elm. The hole is dug out by both birds, and is made on a very similar plan to that adopted by the Green Woodpecker, only the diameter of the shaft is not more than half as much. At the bottom of the hole, which may be but six inches or as much as a foot in depth, a slight enlargement is made, in which the eggs are deposited. No nest is provided for them, beyond the wood-dust collected at the bottom. In exceptional cases this Woodpecker selects a hole ready-made, or enlarges and alters one to its needs. The bird sits closely, and makes little attempt to leave the hole until lifted from it. It is not in any way demonstrative at the nest, should it only contain eggs.

RANGE OF EGG COLOURATION AND MEASUREMENT: The eggs of the Lesser Spotted Woodpecker are from five to eight in number, six being an average clutch. They are pure white and spotless, smooth, and highly polished. Average measurement, ·76 inch in length, by ·58 in breadth. Incubation, performed by both sexes, lasts fourteen days, on the authority of Thienemann.

DIAGNOSTIC CHARACTERS: The small size, and white, polished appearance distinguish the eggs of this species from those of all others breeding in our islands, likely to be confused with them. The single exception is the egg of the Wryneck, but eggs of that bird are almost constantly larger, heavier, and not so highly polished. Nevertheless, the eggs of the Lesser Spotted Woodpecker require most careful identification, which can only be accomplished *at the nest* with absolute certainty.

Family PICIDÆ. Genus PICUS.

GREAT SPOTTED WOODPECKER.

PICUS MAJOR, *Linnæus*.

Single Brooded, generally.   Laying season, May and June.

BRITISH BREEDING AREA: From Yorkshire southwards the Great Spotted Woodpecker is pretty generally distributed throughout the wooded districts of England and Wales. In the extreme northern counties of England it is rare; whilst in Scotland the evidence of its breeding at the present time is by no means conclusive. It is not known to breed anywhere in Ireland.

BREEDING HABITS: The Great Spotted Woodpecker is a resident in this country, and from its shy and secretive habits one that is apt to be often thought more rare and local than it really is. Its breeding-haunts are much the same as those of the Woodpeckers already described. The bird may be met with in woods, open forests, and parks, especially such as abound with old and more or less decayed timber; in larch and fir plantations, in the fringe of trees by the water-side, and in well-wooded agricultural districts. This Woodpecker appears also to pair for life, yet only lives in close company with its mate during the breeding season. Season by season, in many cases, the same nesting-place is tenanted. The nest-hole is generally dug out of a branch or tree-trunk by the parent birds, but occasionally a cavity already made is taken possession of, and enlarged or not according to circumstances. This shaft is made in a precisely similar manner to that of the other species, descending from a few inches to a foot or more into the timber, and being somewhat enlarged at the bottom into a chamber, where the eggs are laid. No nest of any kind is made, and the eggs lie on the soft powdered wood at the bottom of the shaft. The birds are close sitters, undemonstrative, and usually make no attempt to leave their charge until lifted from the eggs.

RANGE OF EGG COLOURATION AND MEASUREMENT: The eggs of the Great Spotted Woodpecker are from five to eight in number. They are white, with a faint yellowish or creamy tinge, and entirely spotless. Average measurement, 1·0 inch in length, by ·78 inch in breadth. Incubation, performed by both sexes, lasts from fourteen to sixteen days.

DIAGNOSTIC CHARACTERS: The eggs of the Great Spotted Woodpecker are best distinguished by their size and creamy tinge. They cannot readily be confused with those of the other Woodpeckers breeding in our islands. Those of the Dipper most nearly resemble them, but the rougher texture and unpolished surface, together with the very different nest, prevent much chance of confusion.

Family CUCULIDÆ.   Genus CUCULUS.

CUCKOO.

CUCULUS CANORUS, *Linnæus.*

Single Brooded. Laying season, end of May to beginning of July.

BRITISH BREEDING AREA: The Cuckoo breeds more or less commonly throughout the British Islands, including the Hebrides, the Orkneys, and the Shetlands.

BREEDING HABITS: The Cuckoo arrives in the southern portions of our islands about the middle of April, but is nearly a fortnight later in the more northern districts. It may be met with in every kind of haunt, from the low-lying broads, the woodlands, and rich agricultural districts, to the wild, upland moorlands and even the rocky, wind-swept summits of the mountains. Wherever small insectivorous birds are breeding the Cuckoo may almost with certainty be found. I am of opinion that the Cuckoo pairs annually, each pair remaining together until the complement of eggs is deposited. I am fully aware that this statement is contrary to the expressed opinions of other naturalists; but it is based upon much careful observation; whilst many of the remarks and opinions published on the matter of the Cuckoo's sexual instincts are unmitigated nonsense. It is said that the birds return yearly to the same locality, but of this I can say nothing from experience. The Cuckoo does not make any nest, but lays her egg upon the ground, and then carries it in her mouth to the nest of some small insectivorous bird who plays the part of foster-parent to her offspring. The eggs of the Cuckoo have been found in a great variety of nests, usually an open one, and those of the Hedge Accentor,

Meadow Pipit, and Reed Warbler perhaps by preference. Less frequently a domed nest like the Wren's, or a nest in a covered site like the Pied Wagtail's or the Redstart's is selected. Usually the nest belonging to an insectivorous species is chosen; more rarely that of a Finch, a Bunting, or even a Red-backed Shrike. Exceptionally the eggs of the Cuckoo have been found in such very ill-adapted nests as those of the Jay, the Magpie, and the Little Grebe. There can be little doubt that the Cuckoo will drop her egg into the nest of a great variety of birds, although few of them are ever discovered by naturalists.

RANGE OF EGG COLOURATION AND MEASUREMENT: The number of eggs laid on an average by each individual female Cuckoo every season is difficult to discover, but probably the estimate of from five to eight is tolerably correct. Two and even three eggs have been found in one nest, but whether the produce of the same female is somewhat problematical. They vary considerably in colour, and as they frequently, but by no means generally, resemble those in the selected nest in this respect, it has been maintained that the female Cuckoo possesses the power of laying eggs similar in tint to those of the chosen foster-parent at will. The eggs of the Cuckoo, however, appear to present several well-marked types, like the eggs of the Tree Pipit, the Red-backed Shrike, and many other birds, each type probably not subject to a very great amount of variation, although certainly to some. The Cuckoos laying eggs of each of these certain types, select as far as possible the nests of species containing eggs most similar to them in colour; whilst the young Cuckoos, with an inherited tendency to produce eggs like their parent, must have a strong inclination to seek the nest of those birds that served as their own foster-parents when the time comes for them to lay eggs of their own, and to provide for their ultimate maturity. It is almost impossible, with the small amount of space at our disposal, to give an adequate idea of the range of colouration of the eggs of the Cuckoo. Although sometimes met with entirely blue and spotless (like a big egg of a Redstart or of a Hedge Accentor), the more usual ground colour varies from grayish-white to greenish-white, spotted, blotched, and freckled with various shades of olive-brown or reddish-brown, intermingled with minute specks of dark brown. Some varieties closely resemble those of the Sky-lark; others those of the Pied Wagtail, the Reed Warbler, or the Meadow Pipit. Average measurement, ·85 inch in length by ·75 inch in breadth. Incubation, performed by the foster-parent, lasts about fourteen days.

DIAGNOSTIC CHARACTERS : It is simply impossible to give any character by which the eggs of the Cuckoo may be identified. They so closely resemble the colour of those of other species, that no reliable distinction is presented, although it may be remarked, the small, round, nearly black specks are very characteristic of the eggs of the Cuckoo. Their size again is unreliable, for we are always confronted with the possibility that instead of a Cuckoo's egg it is merely an abnormally large variety of that of the foster-parent. When placed in nests where the eggs are very different in colour, however, there can be little difficulty.

Family CYPSELIDAE.   Genus CYPSELUS.

COMMON SWIFT.

CYPSELUS APUS, *Linnæus*.

Single Brooded.   Laying season, May and early June.

BRITISH BREEDING AREA: The Swift is generally distributed throughout most parts of the mainland of the British Islands, becoming rarer in the extreme north of Scotland, and in the south-west of Ireland.

BREEDING HABITS: The Swift must be classed amongst the latest of our summer migrants, not arriving in the southern districts until the end of April, and a week or more later still in the north. Like the ubiquitous House Martin, the Swift is almost cosmopolitan, and may be met with breeding amongst all kinds of scenery, from the rocky coast to the upland and mountain districts. It is most abundant, however, in the well-cultivated and more thickly populated areas, breeding freely in villages and in quiet country towns and cathedral cities. The Swift unquestionably pairs for life, and yearly returns to its old nesting-places. These are most generally situated in holes in buildings, especially in church towers and cathedrals, and under thatch or eaves of buildings of all kinds. Sometimes a hole in the wall is chosen; in less populated districts holes in maritime cliffs, in the sides of quarries, and inland in crevices of rocks. Less frequently a hole in the trunk or the branch of a tree is selected. The holes are at varying heights, little choice being apparent, and the bird seems as contented with a hole in the thatch of a low-roofed cottage, almost within arm's-reach, as with one in the lofty walls of more noble edifices. The nest scarcely deserves the name, being merely a few straws or grass-bents lined with feathers, the whole sometimes being matted together with viscid saliva. In many cases the old nest of a Sparrow or of a Starling will be utilized, or the materials annexed. I do not think that the Swift voluntarily glues its nest together with saliva, but in conveying the materials and arranging them they get more or less coated with the abundant secretion. Swifts breed in colonies of varying size according to the amount of accommodation afforded. Sometimes odd pairs may be met with breeding isolated here and there; sometimes the nests are far apart, even in the same building or village, but all through the summer the birds themselves continue gregarious.

RANGE OF EGG COLOURATION AND MEASUREMENT: The eggs of the Swift normally are two in number, but occasionally three and even four have been found. They are white and spotless, very narrow and elongated, rough in texture, and with little or no polish. Average measurement, 1·0 inch in length by ·65 inch in breadth. Incubation, performed by both sexes, lasts seventeen or eighteen days.

DIAGNOSTIC CHARACTERS: The size, white colour, and elongated shape, readily distinguish the eggs of the Swift from those of all other birds breeding in our islands.

Family CAPRIMULGIDÆ. Genus CAPRIMULGUS.

COMMON NIGHTJAR.
CAPRIMULGUS EUROPÆUS, *Linnæus*.

Single Brooded. Laying season, May and June.

BRITISH BREEDING AREA: In districts suited to its requirements the Nightjar is very widely distributed throughout the British Islands, with the exception of the Outer Hebrides, the Orkneys, and the Shetlands. It is certainly most abundant in the southern counties of England, and in Ireland is commonest in the central and southern districts.

BREEDING HABITS: The Nightjar is one of the very latest of our summer migrants, not reaching the British Islands until nearly the middle of May. Its breeding-haunts are on commons and heaths, the rough bracken- and bramble-clothed outskirts of forests, open places in large woods, and spruce and fir plantations, large gorse coverts, and the vicinity of old sand-pits and stone-quarries. It is most probable that the Nightjar pairs for life, and yearly resorts to one particular place to breed. It is not in any sense a gregarious bird, although in suitable localities several may often be seen in the air together, and a few pairs frequently breed within a small area of ground. The Nightjar makes no nest of any kind, the eggs being laid on the bare ground, sometimes on the fallen trunk of a tree, or at the foot of a gorse-bush, or even on stone-heaps. Frequently enough the eggs are found resting in a little hollow, but this is either selected ready-made, or formed by the constant sitting of the brooding bird in one spot. The Nightjar is a remarkably close sitter, as is the universal custom of all birds whose plumage is of a protective hue, remaining brooding over the eggs until almost trodden upon, and then often displaying various alluring antics to entice an intruder from the spot.

RANGE OF EGG COLOURATION AND MEASUREMENT: The eggs of the Nightjar are only two in number, elongated and oval in shape, rounded equally at both ends. They vary from pure white to white with a creamy tinge in ground colour, mottled, blotched, veined, streaked, clouded, and marbled with various shades of brown, and with underlying markings similar in character, of violet gray. Considerable variation, not only in the shade of colour and style of marking, but in the amount of surface-spots and underlying ones occurs. One variety has few of these gray markings, and has the surface-spots numerous and large; another variety has the gray markings large, numerous, and conspicuous, and the surface-spots few and small; another variety is beautifully lined and marbled with brown and blotched with gray; another in which the gray markings are reversed. Average measurement, 1·3 inch in length, by ·87 inch in breadth. Incubation, performed chiefly by the female, lasts from fifteen to eighteen days.

DIAGNOSTIC CHARACTERS: The size, oval shape, and peculiar veined character of the brown and gray markings, readily distinguish the eggs of the Nightjar from those of any other bird breeding in the British Islands.

Family UPUPIDÆ.   Genus UPUPA.

HOOPOE.

UPUPA EPOPS, *Linnæus*.

Single Brooded.   Laying season, latter half of May and first half of June.

BRITISH BREEDING AREA: Were it not for the senseless practice of shooting every rare bird that chances to visit our islands, either on normal passage or otherwise, there can be no doubt that the Hoopoe would, like the Golden Oriole, soon become a regular breeding species in them. Instances of the Hoopoe breeding in Kent, Sussex, Surrey, Hampshire, Wiltshire, Dorset, and Devonshire are on record—a melancholy indication of the way this handsome bird has endeavoured, from one end of our southern coast-line to the other, to establish itself and to spread northwards over the British area.

BREEDING HABITS: I have seen nothing of the breeding habits of the Hoopoe in this country, but in Algeria the bird is common enough, and I met with it in various parts of my wanderings about that charming portion of Africa. The Hoopoe is a summer migrant to Europe, and usually reaches the British Islands in April. Its haunts in Europe are agricultural districts, the outskirts of woods, and in the more open parts of old forest lands; in Algeria, however, I saw this species in the wildest districts, far up the sterile mountain sides, as well as amongst the luxuriant vegetation of the oases. It probably pairs for life, and the nest is made in a hole in a tree, in a crevice of a rock, or in a hole in a wall. This hole is never excavated by the birds. In China, a hole in an exposed coffin is sometimes tenanted; whilst the bird has even been known to make its nest in the chest of a decaying corpse. The nest is merely a few bits of dry grass, straw, or roots, more or less mixed with offensive matter of some kind. The nest often emits a terrible stench, owing to this proclivity, and the fact that the droppings are allowed to remain in the hole. The eggs, however, sometimes rest on the powdered wood alone. The Hoopoe is a close sitter, allowing herself to be lifted from the eggs rather than desert her charge.

RANGE OF EGG COLOURATION AND MEASUREMENT: The eggs of the Hoopoe are from five to seven in number, pale greenish-blue, or pale olive, lavender-gray, or even pale buff in colour, without markings. The shell is more or less glossy, and full of minute pits and streaky hollows, something like the rind of a melon on a very small scale, and which often gives them an appearance of being dusted over with white specks. Average measurement, 1·0 inch in length, by ·7 inch in breadth. Incubation, performed by the female, is said by Naumann to last sixteen days.

DIAGNOSTIC CHARACTERS: The remarkable shell-texture of the eggs of the Hoopoe readily distinguishes them from the eggs of all other birds breeding in our islands, or even in Europe.

Family ALCEDINIDÆ. Genus ALCEDO.

COMMON KINGFISHER.

ALCEDO ISPIDA, *Linnæus.*

Single Brooded. Laying season, rather irregular, April, May, June.

BRITISH BREEDING AREA: Due allowance being made for locality the Kingfisher is more or less sparingly distributed throughout the British Islands, with the exception of the Hebrides, the extreme northern portions of the Scottish mainland, the Orkneys, and the Shetlands. It cannot be regarded as abundant anywhere, and in most localities has decreased and continues to decrease in numbers, owing to senseless persecution.

BREEDING HABITS: The Kingfisher is a resident in our islands. Its favourite breeding-haunts are the fairly well wooded streams and rivers, and the banks of secluded ponds and lakes. I am of opinion that the Kingfisher pairs for life, although the sexes do not keep very close company, except during the nesting season. It is not a gregarious nor even a social species, each bird, or at most each pair, having a vested right in a length of the stream, or the neighbourhood of a pond, to which they keep pretty closely. The nest of the Kingfisher is placed in a hole, usually in the bank of a stream or pool, but now and then in the sides of a gravel- or sand-pit, some distance from the water. Less frequently still a hole in a wall is chosen. In some cases a rat-hole is selected, in others the birds themselves burrow one. In many cases the entrance is concealed either by drooping branches or exposed roots, one of which usually serves as a perching-place for the birds. Occasionally a hole will be made in the same bank that contains a colony of Sand Martins.

Very often a few droppings at the entrance or on the branches near betray the secret of the nest. The hole, which takes about a fortnight to complete, is from three to four feet in depth, slopes upwards, and enlarges into a chamber at the end. Here, often on the refuse of the bird's food, and generally a varying amount of fish-bones, with no other nest whatever, the eggs are deposited. The Kingfisher sits closely, and is a remarkably shy and wary bird at the nest.

RANGE OF EGG COLOURATION AND MEASUREMENT: The eggs of the Kingfisher are from six to ten in number, eight being an average clutch. They are pure and spotless white, with a considerable amount of polish, and very globular in shape. Average measurement, ·9 inch in length, by ·75 inch in breadth. Incubation, performed by both sexes, lasts from fourteen to sixteen days.

DIAGNOSTIC CHARACTERS: The pure and glossy white colour, combined with the rotund form, readily distinguish the eggs of the Kingfisher from those of any other species breeding in our islands.

Family STRIGIDÆ. Genus ALUCO.
Sub-family *STRIGINÆ*.

BARN OWL.

ALUCO FLAMMEUS (*Linnæus*).

Double Brooded. Laying season, April to October, and exceptionally later.

BRITISH BREEDING AREA: The Barn Owl is pretty generally distributed throughout the British Islands, with the exception of the Outer Hebrides, the Orkneys, and the Shetlands. It becomes much more local and scarce in the Highlands, probably owing to the absence of suitable haunts.

BREEDING HABITS : This, the most familiar of all our British Owls, is a resident in the British Islands, and confined for the most part to well-cultivated and populated districts. It frequents ruins, outbuildings, farmhouses, church towers, dove-cots, hollow trees, and ranges of cliffs; less frequently pine woods. The Barn Owl pairs for life, and lives in close company with its mate for the greater part of the year. The bird is neither gregarious nor social, and each pair keep to a chosen haunt and to themselves. The day haunt of the Barn Owl is usually the bird's nesting-place. This hole is usually thickly strewn with pellets containing the refuse of the bird's prey; and on these pellets the eggs are laid, with no other provision whatever. The bird keeps close to its nest until compelled reluctantly to leave it.

RANGE OF EGG COLOURATION AND MEASUREMENT : The eggs of the Barn Owl are from three to six, or rarely seven, in number. They are somewhat rough in texture, with no polish, and pure white without markings. Average measurement, 1·6 inch in length, by 1·2 inch in breadth. Incubation, performed by both sexes in turn, lasts about twenty-one days. In many cases the eggs are sat upon as soon as laid, so that eggs in various stages of development, and chicks of different ages may be found in the same nest together.

DIAGNOSTIC CHARACTERS : The situation of the "nest" is the best means of distinguishing the eggs of the Barn Owl from those of the Long-eared and Short-eared Owls; whilst from those of the Tawny Owl (nesting in similar situations) they are readily distinguished by their smaller size.

Family STRIGIDÆ. Genus STRIX.
Sub-family BUBONINÆ.

WOOD OWL.

STRIX ALUCO, *Gerini.*

Double Brooded.   Laying season, March to July or August.

BRITISH BREEDING AREA: The Wood Owl, otherwise known as the Brown or Tawny Owl, is a fairly well distributed species over England and Wales, although everywhere decreasing in numbers, thanks to the gamekeeper. In Scotland it can only be said to be at all frequent in the southern districts, becoming rare and local in the more treeless areas of the north, although it ranges as far north as Caithness and as far west as Skye and some other of the Inner Hebrides. This Owl appears to be entirely absent from Ireland.

BREEDING HABITS: The Wood Owl is a resident species, and confined for the most part to extensive woodlands, either of pines, firs, or deciduous trees. It is particularly attached to old forest districts and parks where hollow and decayed timber is abundant. More rarely it frequents caves, ruins, barns, and farm-buildings. I am of opinion that this Owl pairs for life, and in many cases the same nesting-place is used year by year. It is neither gregarious nor social in its habits, each pair keeping entirely to themselves. The young are usually reared in the hole which forms the birds' retreat during the day, but not always, as instances are not rare in which the breeding-place is only used for that purpose. The nest-hole is generally in a hollow tree of some kind, but occasionally a hole in a building or a rift in a cliff is selected. Deserted nests of Hawks, Magpies, and Carrion Crows, and old squirrels' dreys are also used at times, whilst a rabbit-burrow or fox's hole is a not very exceptional site. Instances are on record of this Owl breeding in a hen-house, and even in a disused dog-kennel, lying on a lawn within thirty yards of a house! The eggs have also been found on the bare ground under the branches of a fir tree. The Wood Owl makes no nest, and the eggs are usually laid on an accumulation of pellets or whatever may chance to be in the hole —dry earth, decayed wood, dust, etc. The bird is a close sitter, and often resents intrusion in a more or less bellicose manner.

RANGE OF EGG COLOURATION AND MEASUREMENT: The eggs of the Wood Owl are three or four in number. They are pure white, without markings, rotund in form, somewhat smooth in texture, and slightly polished. Average measurement, 1·8 inch in length, by 1·5 inch in breadth. Incubation, performed by both sexes, is said by Naumann to last twenty-one days.

DIAGNOSTIC CHARACTERS: The eggs of the Wood Owl are readily distinguished from those of all other species of Owl breeding in the British Islands by their size—they are the largest of any.

Family STRIGIDÆ. Genus ASIO.
Sub-family *BUBONINÆ*.

SHORT-EARED OWL.

ASIO BRACHYOTUS (*Forster*).

Single Brooded. Laying season, April and May.

BRITISH BREEDING AREA: A few Short-eared Owls still continue to breed in the fen counties of East Anglia, but in the north of England and throughout the wilder districts of Scotland, including the Hebrides, the Orkneys, and the Shetlands, the bird becomes of much more general occurrence. It is not, however, yet known to breed in Ireland, although well known there as a winter visitor.

BREEDING HABITS: It is impossible to say whether the individuals of this species that breed in our islands draw southwards in autumn or not, as at that season great numbers visit our shores from the Continent to winter. The Short-eared Owl is not a dweller in wooded districts, neither does it frequent buildings or cliffs. Its haunts are the open, treeless country, moorlands, fens, heaths, gorse coverts, and marshy wastes. Although several nests may be found at no great distance apart on suitable ground, this Owl is neither gregarious nor social during the breeding season. It pairs very probably for life, and what is very remarkable for a bird laying white eggs, it nests in the open on the bare ground. The nest is usually made amongst sedge or heather or stunted willows, or under the shelter of gorse, or even on a heap of mown reeds. The structure is slight, little more than a hollow lined with a few bits of dry vegetation, and by chance a feather or two. Sometimes when disturbed from the nest, this owl will soar into the air, and hovering above the intruder in company with her mate, watch his movements; at other times, although sitting closely, she flies quite away without further demonstration.

RANGE OF EGG COLOURATION AND MEASUREMENT: The eggs of the Short-eared Owl are from five to seven or even eight in number, smooth, with little polish, and creamy-white in colour. Average measurement, 1·6 inch in length, by 1·3 inch in breadth. Incubation, performed probably by both sexes, lasts from twenty-one to twenty-five days.

DIAGNOSTIC CHARACTERS: The pale creamy tint on the eggs of the Short-eared Owl readily distinguishes them from those of the Long-eared Owl, which are pure white. The nesting-sites of the two birds, however, are also very dissimilar. They cannot readily be confused with those of any other species breeding in our islands, except perhaps those of the Stock Dove, but even then the breeding-grounds are usually different.

Family STRIGIDÆ.   Genus Asio.
Sub-family BUBONINÆ.

LONG-EARED OWL.

Asio otus, *Linnæus.*

Single Brooded.   Laying season, end of February to beginning of April.

BRITISH BREEDING AREA: With the exception of the Orkneys and the Shetlands, the Long-eared Owl is pretty generally distributed throughout the British Islands, its habitat to a great extent being influenced by the presence of pine and fir woods, to which the bird is much addicted.

BREEDING HABITS: The Long-eared Owl is a resident in our islands, and in many districts a fairly common bird, in spite of the persecution of the gamekeeper. Here in Devonshire, for instance, this fine Owl not many years ago was abundant in the fir woods near Churston; now it is scarce, the keeper having destroyed almost every bird. The favourite haunts of the Long-eared Owl are pine woods and fir plantations, or large woods of deciduous trees in which evergreens are intermingled.

It is also partial to the fir plantations on hill-sides and near moors and commons. The Long-eared Owl is not gregarious nor social during the breeding season, but numbers of pairs will breed in a comparatively small area of suitable wood. This Owl does not make any nest, but takes possession of the deserted home of a Magpie, Crow, Wood Pigeon, or Heron, or the old drey of a squirrel. These nests appear not to undergo alteration of any kind. As a rule nests are preferred in such trees that have plenty of branches below them, and at a height of some twenty or thirty feet from the ground. The bird sits remarkably closely, not leaving the eggs until the nest is almost reached.

RANGE OF EGG COLOURATION AND MEASUREMENT: The eggs of the Long-eared Owl are from three to six in number, sometimes seven. They are rather oval in shape, somewhat polished, and pure white. Average measurement, 1·6 inch in length, by 1·3 inch in breadth. Incubation, performed by both sexes, lasts about twenty-seven days.

DIAGNOSTIC CHARACTERS: The pure white eggs of this Owl are easily distinguished from those of the Short-eared Owl, but not so readily from those of the Ring Dove; and as the latter bird breeds in the same woods and plantations they require most careful identification.

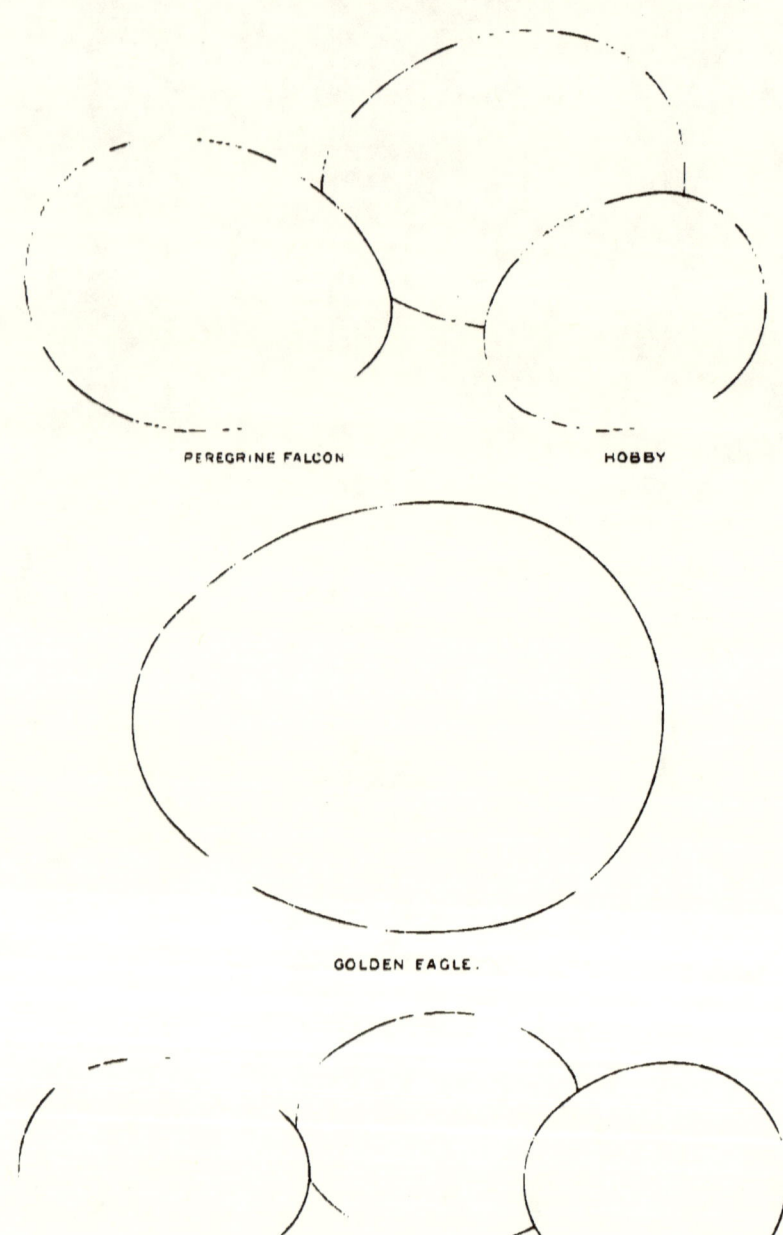

Family FALCONIDÆ. Genus FALCO.
Sub-family FALCONINÆ.

PEREGRINE FALCON.

FALCO PEREGRINUS (*Gerini*).

Single Brooded. Laying season, April and early May.

BRITISH BREEDING AREA: The Peregrine Falcon still breeds locally on the maritime cliffs of England and Wales, and even inland on high rocks in the wilder districts of the north of England—in Cumberland and Westmoreland. In Scotland it becomes more plentiful, and may be said to be pretty generally distributed in all suitable districts, extending to the Hebrides and St. Kilda on the west, and to the Orkneys and Shetlands on the north. In Ireland the bird is equally widely distributed; in some parts even fairly common.

BREEDING HABITS: The Peregrine is a resident in our islands. Its haunts are the bare, treeless districts, the long ranges of ocean cliffs, wild mountain sides, and breezy moorlands. This bold Falcon, in my opinion, pairs for life, and if it does not always frequent the same nest every year, it usually returns to one favourite spot to breed, often using several sites in turn. Certain cliffs have been known to contain a Peregrine's eyrie from time immemorial. The Peregrine is not a social species, each pair confining themselves to certain districts, or beats, which they appear to guard jealously from intrusion. They are also much attached to particular haunts, and will return to breed in them year by year in spite of continued disturbance. The nest (if such it can be called) is usually on a narrow ledge of a cliff, frequently where the rocks overhang. The Peregrine cannot be said to make a nest, the hollow in the soil on the cliffs in which the eggs are deposited being selected ready to hand, or made undesignedly by the old birds whilst incubating. At a nest on the Bass, a few scraps of heather were in the hollow, as well as a feather or so; but these materials had most probably accumulated there by chance. The Peregrine is not a very close sitter, and when disturbed, often flies to and fro high in the air overhead, chattering loudly in displeasure at the invasion of her fastness. The male bird generally soon joins her.

RANGE OF EGG COLOURATION AND MEASUREMENT: The eggs of the Peregrine are from three to four in number, but sometimes only two, and I have known a nest that contained one solitary chick. They are yellowish or creamy-white in ground colour, thickly mottled, spotted, and clouded with reddish-brown, brick-red, and orange-brown of various shades. The markings are usually so numerous and so confluent, that little if any of the ground colour is visible. A rare variety is suffused with a purplish tinge or bloom. These eggs closely resemble those of the Kestrel in colour, and go through precisely the same variations. Average measurement, 2·0 inches in length, by 1·6 inch in breadth. Incubation, performed chiefly by the female, lasts a month according to Macgillivray.

DIAGNOSTIC CHARACTERS: The large size of the eggs of this Falcon readily distinguish them from those of all allied species breeding in the British Islands.

Family FALCONIDÆ. Genus FALCO.
Sub-family FALCONINÆ.

HOBBY.

FALCO SUBBUTEO, *Linnæus*.

Single Brooded. Laying season, latter half of May or first half of June.

BRITISH BREEDING AREA: The Hobby is one of the rarest and most local of the Falcons that breed in our area. It breeds very sparingly in the south-eastern and midland counties of England, including Hampshire, Essex, Suffolk, Norfolk, Cambridgeshire, Lincolnshire, Leicestershire, Northamptonshire, Derbyshire, and Yorkshire. Not known to breed in Scotland or Ireland.

BREEDING HABITS: The Hobby is a regular summer migrant to the British Islands, reaching them towards the end of April or during the first half of May. The favourite breeding-grounds of this bold little Falcon are the large woods and plantations. Everywhere it is a forest bird, and never appears to breed in any other localities. It is most probable that the Hobby pairs for life, and, like its congener the Merlin, returns year by year to certain spots to breed. In fact, many of these breeding-places are so well recognized, that instances are on record where pair after pair of birds have been shot at them for many years in succession. The Hobby is not at all social during the breeding season, but several pairs may be met with breeding in one vicinity. The Hobby, like the Kestrel, is no nest-builder, but chooses the deserted home of a Crow, a Magpie, or even a Wood Pigeon, in which to lay its eggs and bring them to maturity. A nest in a tall tree in a secluded part of the woods is usually selected. The nest appears not to undergo any alterations in our islands; but in Germany the bird is said to re-line the selected nest with wool and feathers. I should say that in some cases, especially where a Magpie's nest is chosen, the lining of roots is removed. When disturbed at the nest both birds become more or less demonstrative, circling above and round the tree, and uttering shrill chattering cries of alarm.

RANGE OF EGG COLOURATION AND MEASUREMENT: The eggs of the Hobby are usually three in number, but Lord Lilford records instances of four young birds being taken from one nest; so that it would appear that this number is sometimes exceeded. They are yellowish-white in ground colour, freckled, mottled, suffused, and blotched with reddish-brown. On most eggs little of the ground colour is visible, and they go through every variation that is presented in the eggs of the Kestrel. Average measurement, 1·7 inch in length, by 1·3 inch in breadth. Incubation is said to last three weeks, and to be performed by both sexes.

DIAGNOSTIC CHARACTERS: It is impossible to give any character by which the eggs of the Hobby may be distinguished from those of the Kestrel and the Merlin. Nothing but the most careful identification at the nest is of any use in authenticating them.

Family FALCONIDÆ. Genus FALCO.
Sub-family *FALCONINÆ*.

MERLIN.

FALCO ÆSALON (*Brisson*).

Single Brooded.   Laying season, May.

BRITISH BREEDING AREA: The Merlin only breeds very locally south of the Peak in Derbyshire, on Exmoor, and in Wales. From the southern spurs of the Pennine Chain it becomes more widely and generally distributed throughout the moorland districts of England and Scotland, west to the Outer Hebrides, and north to the Orkneys and Shetlands. In Ireland it is equally common in the moorland districts.

BREEDING HABITS: The Merlin is a regular bird of passage, in the sense that it leaves its upland breeding-haunts in autumn for the more low-lying districts, returning to them in April. As the favourite breeding-places are tenanted yearly, there can be little doubt that the Merlin pairs for life; it has also repeatedly been remarked that in places where the birds have been shot season after season, a new pair of birds take possession of the favourite site. The Merlin is not a social bird, but several nests may not unfrequently be met with on a comparatively small area of ground. The breeding-haunts of this Falcon are on the heathery moors, the haunt of the Red Grouse, the Twite, and the Ring Ouzel. The site for the nest is usually on the rough ground among heather and ling, a spot on a sloping eminence commanding a good look-out, or on the rock-strewn earth at the foot of a low range of cliffs. In Scotland, however, the Merlin is said occasionally to select an old nest of some other bird in a tree—but this must be very exceptional—whilst in the Faroes and some other countries a ledge on the cliffs is chosen. But little if any nest is made; sometimes a mere hollow, sometimes a slight ring of twigs. The Merlin is not a very close sitter, and when flushed occasionally flies round in circles uttering a low, chattering cry, especially if the eggs be much incubated.

RANGE OF EGG COLOURATION AND MEASUREMENT: The eggs of the Merlin are four or five in number, very rarely six. They are creamy-white in ground colour, clouded, mottled, and spotted with reddish-brown, and occasionally marked with a few very dark brown specks. Usually the eggs are so richly marked that little or none of the ground colour is visible; in some the markings are purplish-red; in others the ground colour is more apparent, and the markings are small and dark and dusted over most of the surface. The eggs of this Falcon run through every type that is to be seen in the eggs of the Kestrel; they are rarely banded or zoned. Average measurement, 1·6 inch in length by 1·2 inch in breadth. Incubation, performed by both sexes, lasts four weeks.

DIAGNOSTIC CHARACTERS: The eggs of the Merlin cannot be distinguished from those of the Kestrel or the Hobby: the colour is perhaps on an average browner, not so brick a red. The situation of the nest, however, prevents much possibility of confusion.

Family FALCONIDÆ. Genus FALCO.
Sub-family *FALCONINÆ.*

KESTREL.

FALCO TINNUCULUS, *Linnæus.*

Single Brooded. Laying season, latter end of April or in May.

BRITISH BREEDING AREA: The Kestrel is the commonest and most widely distributed of the British Birds of Prey. It breeds and is generally dispersed throughout the British Islands on uplands and lowlands alike, in the wildest as well as in the well-cultivated districts.

BREEDING HABITS: In many districts the Kestrel is a regular migrant, leaving in autumn and returning in early spring (March); still in some of our southern counties individuals may be met with all the winter through, although it is difficult to say whether these are not migrants from the Continent, and that our own breeding birds uniformly winter further south. The Kestrel is not very particular as to the choice of a breeding haunt. It may be met with breeding almost anywhere, from the maritime cliffs to the inland woods, from the ivied ruin or outbuilding to the moorland or mountain plantation or range of rocks. No one area seems more favoured than the other. The Kestrel is not a social species, although several nests may be met with at no great distance from each other. I am of opinion that this bird pairs for life, using the same place season after season to rear its young. The Kestrel is no nest-builder. When breeding on the rocks its eggs are laid on the bare earth of some ledge or crevice, when in a hole in a building on the bare dusty masonry, when in a hollow tree on the decayed and powdered wood at the bottom of the cavity. Many Kestrels, however, make use of a deserted nest of a Crow, a Magpie, a Sparrow-Hawk, or even a Ring-Dove. When a Magpie's nest is chosen, I have invariably found that the lining of roots has been removed. The eggs are usually found resting on or are surrounded by a varying number of pellets of food refuse cast up by the parent Kestrels. The Kestrel is a somewhat close sitter, and when flushed is not very demonstrative unless the eggs are incubated to some extent. Both birds will then fly above the place and chatter mournfully.

RANGE OF EGG COLOURATION AND MEASUREMENT: The eggs of the Kestrel are usually six in number, rarely seven, and sometimes only four or five. They are rotund in shape, and creamy-white in ground colour, washed, spotted, and blotched with reddish-brown. As a rule but little if any of the ground colour is visible. They vary considerably, yet most are richly marked. Rare varieties, however, have most of the colour massed on the larger end; and some instead of having the brown washed over most of the surface in varying intensity, have it broken up into fairly well defined blotches. Average measurement, 1·6 inch in length by 1·3 inch in breadth. Incubation, performed chiefly by the female, lasts about four weeks. Sometimes the eggs are sat upon as soon as laid, and often two days or more elapse between each successive egg.

DIAGNOSTIC CHARACTERS: It is impossible to distinguish the eggs of the Kestrel from those of the Merlin or the Hobby. The situation of the nest as regards the former species will however suffice; but as regards the latter the most careful identification is required.

Family FALCONIDÆ. Genus AQUILA.
Sub-Family AQUILINÆ.

GOLDEN EAGLE.

AQUILA CHRYSAETUS (*Linnæus*).

Single Brooded. Laying season, March or early April.

BRITISH BREEDING AREA: Two hundred years ago the Golden Eagle bred in Derbyshire; less than a hundred years ago its nest might be found in the districts of the English Lakes. It is now banished from every lowland haunt, and only breeds locally and in limited, if slowly increasing, numbers in the remote fastnesses of the Highlands and the Hebrides. Here we are glad to know it is protected to a certain extent by the landowners, and doubtless will for years to come continue to breed. In Ireland also it was formerly a fairly common bird; now but a few odd pairs survive in the wilder districts of the north and west.

BREEDING HABITS: The Golden Eagle is a resident in our islands, but wanders about a good deal during the non-breeding season. Its nesting haunts are among the most secluded mountains, in the wild romantic glens and near the deep sea lochs, or in the quiet fastnesses of the Highland deer-forests. The Golden Eagle pairs for life, and usually frequents one favourite breeding-place for years, although the same nest may not be used each successive season. The bird is not at all social, only living in close company with its mate during the nesting period, each pair having a sort of vested right in a very wide area of country. The nest is usually placed on some inland crag, often on the side of a glen, and rarely if ever on a tree, in our islands nowadays. The only nest I ever knew on a *maritime* cliff was on the west coast of Skye, a precipice quite six hundred feet high, partly a broken slope covered with grass and other herbage, and partly crags which fell sheer down almost like a wall. The nest was made in a small fissure, beneath an overhanging piece of the rock, and was quite inaccessible without the aid of a rope. The nest of this Eagle is a massive bulky pile of sticks and branches, generally well interwoven, and lined with dry grass, dead fern leaves and lumps of moss, and usually containing tufts of the mountain plant *Luzula sylvatica*. As a rule this Eagle sits lightly, and if the nest only contains eggs makes little or no demonstration. That it ever attacks a human being at the nest I for one do not believe.

RANGE OF EGG COLOURATION AND MEASUREMENT: The eggs of the Golden Eagle are usually two in number, sometimes three, occasionally only one.[1] They vary from dull white to very pale bluish-green in ground colour, blotched, clouded, spotted, and freckled with reddish-brown, and with underlying markings similar in character of violet-gray. They vary considerably even in the same nest, and it is the exception to find two similarly marked eggs in the same clutch, and very often one or two are much less richly coloured than the rest. Sometimes the spots are pretty evenly distributed over the entire surface, but always most numerous on the larger end of the egg. Another variety is handsomely clouded and blotched with dark reddish-brown, and with most of the space between the large markings dusted with paler brown specks. Sometimes the gray underlying markings are very numerous and large, at other times they are small and few. Another type is almost spotless, and either dirty white or pale bluish-

[1] Four eggs have been recorded as found in one nest, but this must be an extremely exceptional circumstance.

green, with perhaps a few faint rusty markings. Average measurement, 2·9 inches in length by 2·3 inches in breadth. They are usually somewhat rotund in shape, but even in this particular, eggs in the same nest often differ considerably. Incubation, performed by both sexes, lasts about a month. The eggs are sat upon as soon as laid, and often a few days intervene between the deposit of each.

DIAGNOSTIC CHARACTERS: The size and rich markings in most cases readily distinguish the eggs of the Golden Eagle from those of allied species. The colourless varieties may be distinguished from those of the White-tailed Eagle by their fine, almost smooth, texture, those of that species being much coarser grained.

Family FALCONIDÆ. Genus HALIAËTUS.
Sub-family AQUILINÆ.

WHITE-TAILED EAGLE.

HALIAËTUS ALBICILLA (*Brisson*).

Single Brooded. Laying season, March and April.

BRITISH BREEDING AREA: Even less remotely (little more than fifty years) than the Golden Eagle, the White-tailed Eagle bred in the north of England and the south of Scotland; but persecution has now long banished this fine bird from all the more populated areas, and its great stronghold is in the northern districts. It is certainly far more numerous than the Golden Eagle, and may be met with breeding along the wild west coast of Scotland and amongst the Hebrides, reaching as far north as the Shetlands. In Ireland it becomes much less common, although formerly plentiful, and breeds only in a few localities in the wild and secluded western littoral. It has long ceased to breed on the Blasket Isles.

BREEDING HABITS: The White-tailed Eagle is a resident in the British Islands, but wanders about considerably during the non-breeding season. Its breeding-grounds are now almost exclusively confined to the ocean cliffs that occur so frequently on the western coasts of Scotland and in the Hebrides. It occasionally breeds in more inland districts, although never very far from the sea, in localities similar to those selected by the preceding species, or even on an island in a large mountain loch. This Eagle also pairs for life, and returns year by year to breed in one particular spot, to which it is much attached, and which has been the site of an Eagle's eyrie for time out of mind. It may not be any more social than the Golden Eagle, but eyries are frequently found rather close together, and as likely as not a Peregrine's nest may be met with in the same range of cliffs. Many eyries are quite inaccessible, others are placed in comparatively easy situations, which may be climbed to without the aid of a rope. The nest is usually made in some crevice or on a ledge of the cliffs, often sheltered by an overhanging crag. Sometimes it is made in a tree; and instances are on record where a large bush on an island has been chosen. The nest, which is generally patched up and added to each season, is a vast pile of sticks and branches, matted and interlaced, pieces of sea-weed and turf, and lined with dry grass, leaves of the sea-campion and bunches of wool. In some cases, however, the nest is nothing near so elaborate, little more than a hollow in the bare earth on the rock-ledge; whilst the nests that are made in trees are generally the largest. The White-tailed Eagle sits lightly, although if the eggs are much incubated the bird is made to leave them with difficulty. But little demonstration is made at the nest if it only contains eggs, and the stories of Eagles attacking men when disturbed I am disposed to regard as rubbish.

RANGE OF EGG COLOURATION AND MEASUREMENT: The eggs of the White-tailed Eagle are normally two in number, but in very rare instances three have been found, and sometimes only one. They are white, and generally without markings of any kind, although sometimes a few rusty stains occur, but these most probably are accidental. Average measurement, 2·9 inches in length, by 2·2 inches in breadth. They are not quite so bulky as those of the Golden Eagle, being narrower for their length. The incubation period is apparently unknown, but probably about the same as that of the preceding species, and performed by both sexes, principally the female.

DIAGNOSTIC CHARACTERS: The eggs of the White-tailed Eagle may be readily distinguished from those of the Golden Eagle, the only species with which they are likely to be confused in this country, by the absence of colouring matter and their coarse texture.

Family FALCONIDÆ.   Genus MILVUS.
Sub-family *BUTEONINÆ*.

COMMON KITE.

MILVUS REGALIS (*Brisson*).

Single Brooded.   Laying season, May.

BRITISH BREEDING AREA: The Kite is another melancholy relic of our ancient avifauna, once an abundant species even in the streets of London, but now, alas! one of the rarest and most local of our indigenous birds. Twenty-two years ago the last nest was known in Lincolnshire. It still breeds here and there in Wales, and also in a few localities in Scotland which need not be named. The Kite is only an abnormal wanderer to Ireland.

BREEDING HABITS: There can be little doubt that the Kites breeding in our islands are residents. The breeding-haunts of this interesting species are woods and forests; in Scotland the wild secluded pine forests are the special favourite. It is not a social bird, and even keeps closest company with its mate during the breeding season. The nest of the Kite is invariably made in a tree in our islands, but in North Africa a ledge on a rock is sometimes chosen. By preference a dense fir or pine tree is selected, and the nest is built at varying heights from the ground, sometimes as much as forty feet, sometimes only fifteen or twenty feet. It is built either among the more slender topmost branches, or, and more generally, on several branches lower down the stem and close to the trunk. Externally it is made of sticks, branches, and an assorted mixture of rubbish, such as old rags and dirty paper, and lined with moss, wool, and a further selection of odd scraps, including bones, hair, paper, worsted, and rags of all kinds. It is somewhat flat inside, and generally a bulky structure. When disturbed from the nest, which it leaves somewhat reluctantly, the sitting bird often soars high in air above the tree, soon being joined by its mate.

RANGE OF EGG COLOURATION AND MEASUREMENT: The eggs of the Kite are usually three in number, sometimes but two, and very rarely four. They are very pale bluish-green, or grayish-white in ground colour, clouded, blotched, spotted, streaked, and freckled with dark reddish-brown and paler brown, and with underlying markings of violet-gray. The amount of markings varies considerably. Some varieties are clouded and blotched; others spotted and faintly streaked; some have few markings at all, and either most of them underlying or surface ones; on some eggs the markings are small and pretty evenly dispersed; on others large and bold, and mostly confined to the large or small end of the egg. Average measurement, 2·25 inches in length, by 1·8 inch in breadth. Incubation, performed chiefly by the female, lasts, according to Tiedemann, from twenty-one to twenty-four days, but is probably longer.

DIAGNOSTIC CHARACTERS: The eggs of this species may readily be confused with those of the Common Buzzard, and require the most careful identification. The character of the nest, however, renders confusion unlikely.

The Honey Buzzard (*Pernis apivorus*) appears never to have been a very abundant species in the British Islands, although it was certainly known by Willughby to breed in our area two hundred years ago. The high price set upon its eggs has done more than anything else to exterminate this species, and it is now more than doubtful if any individuals rear their young within our area. The New Forest district appears to have been its last lingering stronghold. It may still breed there, but if such is really the case the secret is guarded with a jealousy which every naturalist must respect. The Honey Buzzard does not appear to make its own nest, but to select a deserted nest of a Crow or a Magpie, a Buzzard or a Kite, which it relines with a quantity of green leaves (beech by preference), or twigs with the leaves on them. This lining appears to be renewed from time to time. The eggs are from two to four, generally the former, less frequently three, and very rarely the higher number. They are remarkably handsome, rotund in form, smooth and waxy in texture, cream or pale red in ground colour, boldly blotched and spotted with rich purplish-brown, often so abundantly as to conceal all trace of the ground. Average measurement, 1·95 inch in length, by 1·7 inch in breadth. Incubation is performed by both sexes, and lasts from twenty-one to twenty-three days. Only one brood is reared in the year.

Family FALCONIDÆ. Genus BUTEO.
Sub-family *BUTEONINÆ*.

COMMON BUZZARD.

BUTEO VULGARIS, *Leach.*

Single Brooded. Laying season, April and May.

BRITISH BREEDING AREA: The Common Buzzard breeds very locally on the cliffs of the Welsh coast, and in some of the wooded inland districts of the Principality. A few pairs also nest in some of the large forests of Scotland, and the same remarks I am glad to say still apply to Ireland. It is unwise more specially to indicate the localities, as unfortunately the greed of the collector seldom allows him to respect such confidence.

BREEDING HABITS: The Common Buzzard must be regarded as a resident in our islands, although there is considerable local movement during the non-breeding season. In our islands the breeding-haunts of this Buzzard are either in secluded woods, or in the vicinity of crags and maritime cliffs. The bird is not a social one, although two nests may sometimes be found not

PLATE V

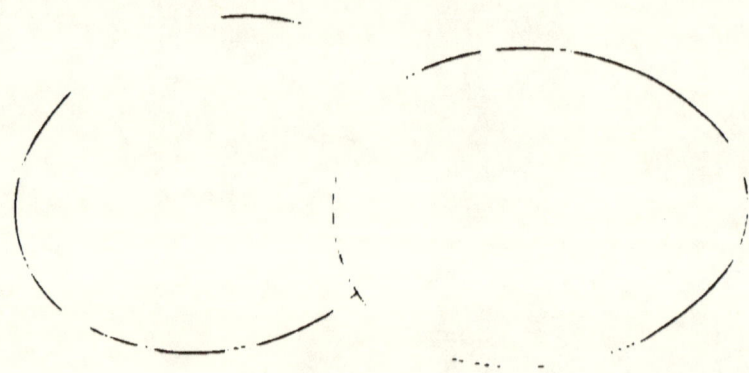

COMMON BUZZARD.

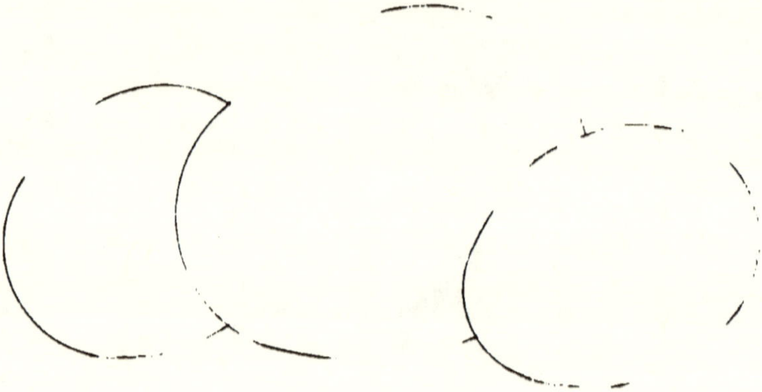

SPARROW HAWK   KITE   SPARROW HAWK

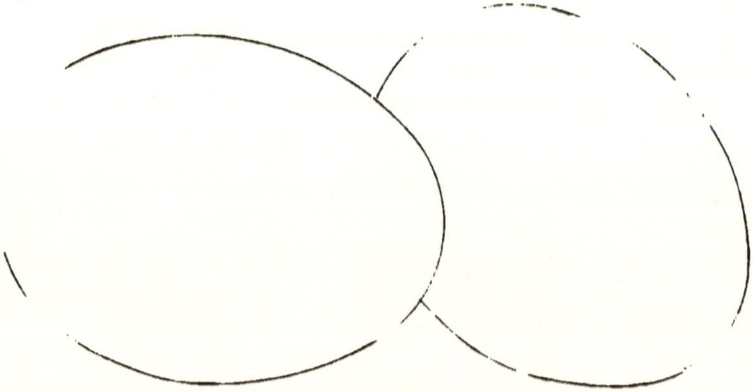

OSPREY.

very far apart, especially in districts where suitable breeding-places are somewhat scarce, or the birds are not molested. This bird most probably pairs for life, and will breed in the same nest season by season if left unmolested. The nest of this species is either made in a tree—a fir or a pine perhaps by preference, in our islands—but in districts where such are not common, as for instance in the New Forest, an oak or a beech will be selected. I have seen it in pines, on a flat branch some distance from the trunk, and (four-and-twenty years ago near Sheffield) up a tall oak in a wood, in a fork close to the main stem. When on cliffs it is often built amongst ivy, or under the shelter of a bush growing from their sides, or in a fissure where the rocks overhang. It is made at various heights, some being ninety or a hundred feet from the ground, others not more than twenty-five feet. The nest is a large, flat, bulky structure, made externally of sticks and twigs, and lined with a scrap or two of down or wool, and a quantity of green leaves. This habit of placing living leaf-decked twigs in the nest is also common to the Sparrow-Hawk, but not quite so apparent. This lining appears to be renewed from time to time until the young are reared, and may be a crude attempt at mimicry to assimilate the structure to surrounding foliage. The bird is a somewhat close sitter, but when disturbed frequently flies round and round above the tree, uttering a monotonous and plaintive cry.

RANGE OF EGG COLOURATION AND MEASUREMENT: The eggs of the Common Buzzard are usually three in number, sometimes two, very rarely four. They vary from white suffused with reddish-brown, to very pale bluish-green in ground colour, blotched, clouded, spotted and streaked with rich reddish-brown, and with underlying markings of violet-gray. They vary a good deal in the character, amount, and distribution of the markings, some being almost spotless, others richly and handsomely clouded and blotched, usually at one or the other end, or less frequently uniformly covered with shorter irregular streaks. A beautiful variety is clouded all over the surface-markings with a thin film of lime, making them look as though they were veiled in a wreath of gray smoke, and giving them a uniform pink appearance. They vary equally as much in form. Average measurement, 2·2 inches in length, by 1·75 inch in breadth. Incubation, performed chiefly by the female, is said to last three weeks.

DIAGNOSTIC CHARACTERS: The eggs of this species cannot readily be confused with those of any other Raptorial bird breeding in the British Islands, except those of the Kite. The latter species, however, is extremely rare, and its nest is a very peculiar one. Nevertheless the eggs should be most carefully identified.

The Rough-legged Buzzard (*Archibuteo lagopus*) is said to have bred regularly in Yorkshire half a century ago, but as the fact is only based on the recollections of a gamekeeper (men of wonderfully elastic memory), I for one place no reliance whatever upon it. The assertion of Mr. Thomas Edward of Banff, that this species had bred in that vicinity, is just as unworthy of credence until more completely substantiated. A description of the nest and eggs and the salient points in the reproductive economy of the Rough-legged Buzzard is therefore quite out of place in a volume dealing with British species alone.

Family FALCONIDÆ. Genus CIRCUS.
Sub-family ACCIPITRINÆ.

MONTAGU'S HARRIER.

CIRCUS CINERACEUS (*Montagu*).

Single Brooded. Laying season, May.

BRITISH BREEDING AREA: Montagu's Harrier still occasionally breeds, or more often attempts to do so, in Norfolk, and instances are on record of its having recently bred in Wales, and some of the southern counties, from Hampshire in the east to Somerset and Devonshire in the west.

BREEDING HABITS: There can be no doubt that Montagu's Harrier is a migratory bird in our islands, visiting them in April, and departing for the south again in October. The few places in which this bird breeds in our islands are principally dry moors and rough heaths, similar localities to those so often chosen by the Stone Curlew. It is probable that this species pairs for life, and regularly returns to certain districts to breed, like the Merlin and the Hobby so frequently do. It is not a social bird, each pair keeping to a particular beat and to themselves. The nest is invariably placed upon the ground on a bare spot amongst the heath, or in a little clearing in a gorse thicket. On the Continent a field of corn is frequently chosen. It is merely a hollow in which a little dry grass or a few straws are arranged, round which are a few twigs. Montagu's Harrier is not a very close sitter, and when disturbed usually flies away in gradually widening circles, returning in the same manner when the disturbance has passed, and dropping suddenly into the nest.

RANGE OF EGG COLOURATION AND MEASUREMENT: The eggs of Montagu's Harrier are from four to six in number. They are pale bluish-white, almost the colour of skimmed milk, and very rarely have a few pale markings of reddish-brown. Average measurement, 1·7 inch in length, by 1·3 inch in breadth. They are laid at intervals of a day or so, and usually the bird begins to sit at once. Incubation, performed by the female, lasts from twenty-one to twenty-four days, according to Tiedemann.

DIAGNOSTIC CHARACTERS: The eggs of this species may be readily distinguished from those of the other two species of British Harrier by their small size; whilst no other bird of prey breeding in our islands (with the above exceptions) on the ground lays eggs at all approaching them in colour.

Family FALCONIDÆ. Genus CIRCUS.
Sub-family *ACCIPITRINÆ*.

HEN HARRIER.

CIRCUS CYANEUS (*Linnæus*).

Single Brooded. Laying season, May and early June.

BRITISH BREEDING AREA: The Hen Harrier is another very local bird, having been almost exterminated from most of its accustomed haunts, especially in England, by the gamekeeper. It still breeds (or attempts to do so) very sparingly and locally on the moors of Cornwall and Devonshire, in Wales, and the lake district. In Scotland its chief strongholds are in the Highlands, the Hebrides, and the Orkneys, although everywhere it appears to be getting rarer. In Ireland it still breeds in some of the mountain districts.

BREEDING HABITS: The Hen Harrier is for the most part a summer visitor to our islands, arriving in April and May, although a few individuals remain in them throughout the winter. The haunts of this rare and conspicuous bird are wild moorlands, and the heath and gorse-clothed sides of mountains. It shows little or no preference for marshy ground, choosing the dry parts of the moors and heaths, and at much higher elevations than those selected by the preceding species. It is not at all social, each pair keeping to a particular area and to themselves. The nest of this Harrier is invariably placed upon the ground, often amongst very tall heather, but sometimes in a more bare and open situation in a little clearing amongst surrounding vegetation. In this country the nest is usually merely a hollow lined with a little dry grass and with a few twigs round the margin, but in some other instances it is a much larger structure a foot or more high, yet made of similar materials. As a rule the smaller nests are made in the barer situations. The bird does not sit very closely, and its actions at the nest are similar to those of the preceding species.

RANGE OF EGG COLOURATION AND MEASUREMENT: The eggs of the Hen Harrier are from four to six in number, five being the usual clutch. They are bluish-white, almost the colour of skimmed milk, and very exceptionally have a few pale rusty markings. Average measurement, 1·8 inch in length, by 1·4 inch in breadth. Incubation, performed by the female, lasts from twenty-one to twenty-four days.

DIAGNOSTIC CHARACTERS: The eggs of the Hen Harrier may usually be distinguished by their size—intermediate between that of the former and the following species—but as exceptionally small examples intergrade with those of the preceding bird, and exceptionally large ones with those of the following, this character is not absolutely trustworthy. Only the most careful identification at the nest is of avail in authenticating them.

Family FALCONIDÆ. Genus CIRCUS.
Sub-family ACCIPITRINÆ.

MARSH HARRIER.

CIRCUS ÆRUGINOSUS (*Linnæus*).

Single Brooded.   Laying season, May.

BRITISH BREEDING AREA: There can be little doubt that the Marsh Harrier will soon cease absolutely to breed in our area if not afforded more protection than is now vouchsafed to it. Probably the only county in which it now regularly breeds or attempts to do so in Great Britain, is in Norfolk. In Ireland it may still breed here and there in suitable districts, but is well-nigh exterminated by the keeper.

BREEDING HABITS: The Marsh Harrier, if left unmolested, is probably a resident in our islands, but undertakes some local movement during the non-breeding season. Its only haunt in our islands is the Broads, the wide wastes of reed and marsh and water that were once the home of so many interesting species. It is too rare in our islands to be social; but in countries elsewhere, this Harrier seems almost to be gregarious during the breeding season; for Irby states that no less than a score of nests have been found within three hundred yards of each other. The nest of this Harrier in our islands is almost invariably made on the ground, but in some countries a low tree is selected. Montagu, from experience in the British Islands, however, records a nest in the fork of a tree. It is usually made amongst reeds or sedge, or under the shelter of a little bush or tuft of herbage. It is a bulky structure composed of reeds, sticks, and twigs, and lined with dry grass and bits of withered aquatic vegetation. In some instances this Harrier probably takes possession of a Coot's nest ; and it is said occasionally to add to the structure during the course of incubation, probably as a protection against sudden floods. The bird is not a very close sitter, usually flying away before the nest is closely approached, and soaring round in circles, joined by its mate, to watch the movements of an intruder.

RANGE OF EGG COLOURATION AND MEASUREMENT: The eggs of the Marsh Harrier are from three to six in number. They are very pale bluish-green, almost white in some cases, very rarely marked sparingly with a few rusty spots. Average measurement, 2·0 inches in length, by 1·5 inch in breadth. They are rotund in shape, and rather rough in texture. Incubation, performed by the female, lasts from twenty-one to twenty-four days, according to Tiedemann.

DIAGNOSTIC CHARACTERS: The large size almost invariably distinguishes the eggs of the Marsh Harrier from those of allied species breeding in our islands. If the measurements of doubtful eggs overlap, it will invariably be found that those of the present bird are more bulky.

Nearly ninety years ago the Goshawk (*Astur palumbarius*) was reputed to breed in the valley of the Spey, but the evidence appears to me to be utterly untrustworthy. That the bird at some remote epoch inhabited the British Islands there can be little doubt; but this was long previous to any ornithological records, and consequently we must decline to admit the species into the present work. As the Goshawk is a bird of very limited migrations, the chance of our islands becoming repeopled with this species is excessively small, for we almost invariably find that when birds become extinct as breeding species with us, they rarely if ever establish themselves again unless introduced by artificial means.

Family FALCONIDÆ. Genus ACCIPITER.
Sub-family ACCIPITRINÆ.

SPARROW-HAWK.

ACCIPITER NISUS (*Linnæus*).

Single Brooded. Laying season, latter end of April and in May.

BRITISH BREEDING AREA: The Sparrow-Hawk, next to the Kestrel, is the most widely distributed of the British birds of prey, and breeds throughout the wooded portions of our islands. In the comparatively bare and treeless districts of the Hebrides and the Orkneys and Shetlands it is rare, absent one might say, as a regular breeding species.

BREEDING HABITS: The Sparrow-Hawk is practically a resident in our islands, although there is a perceptible movement from the wilder and most northerly districts during winter. The haunts of this pretty Hawk are exclusively confined to well-timbered districts, either the woods and coppices of the lowlands, or the spruce and fir plantations of more elevated areas. It is for the most part a solitary species, and although it probably pairs for life, the sexes do not keep very close company, except in the breeding season. In many cases the same nest will be used season by season, or a new one is made each spring in a tree close by that containing the nest of the previous year. The nest of the Sparrow-Hawk is always in a tree of some kind, and always made by the birds themselves. A larch, a pine, or a fir tree are favourite situations in some localities, oaks and alders in others. The nest is rarely made in the slender branches, almost invariably in a crotch where several large limbs spring out, or on the horizontal branch of a larch or pine, close to the trunk. Sometimes it is made at a distance from the trunk, on the dense flat branch of a fir. The nest is large, flat, straggling, and loosely if firmly put together, and made of sticks and twigs; the finest being reserved for the shallow cavity which contains the eggs. Very often a few green branches of the larch are inserted, and bits of down are generally to be seen in the nest and near by. The nest often remains empty for several days when finished. The Sparrow-Hawk is a close sitter, and will often dash by an intruder's head as he climbs up to the nest.

RANGE OF EGG COLOURATION AND MEASUREMENT: The eggs of the Sparrow-Hawk are from four to six in number, five being the average clutch. They are pale greenish-blue in ground colour, spotted, blotched, and clouded with rich reddish-brown, and paler brown. They vary considerably in the amount and richness of the markings, some specimens being so abundantly spotted as to hide most of the ground colour; others are very handsomely zoned round either end, or even round the middle. Rarely they are almost spotless; and occasionally they display one or two large, irregular masses of colour and no more. Average measurement 1·65 inch in length, by 1·3 inch in breadth. Incubation, performed principally by the female, lasts about three weeks. The female will continue to lay egg after egg in the same nest if they are removed. I have known as many as fourteen to be taken from a single nest in one season.

DIAGNOSTIC CHARACTERS: The size of the eggs, pale ground colour, and richness of the markings, readily distinguish the eggs of this species from those of all others breeding in our islands.

Family FALCONIDÆ.   Genus PANDION.
Sub-family PANDIONINÆ.

OSPREY.

PANDION HALIAËTUS (*Linnæus*).

Single Brooded.   Laying season, latter half of April or early in May.

BRITISH BREEDING AREA: The Osprey formerly bred much more commonly in our islands than is now the case. It is one of the rarest and most local of British birds, confined exclusively during the breeding season to one or two favoured spots in the Highlands—in Ross-shire and Inverness-shire.

BREEDING HABITS: The Osprey is a summer migrant to the British Islands, reaching its breeding-grounds in Scotland during May. The favourite haunts of this interesting bird are the wild mountain forests which clothe the heights among which lochs and ruins abound, and the quiet lakes sleep calmly in their setting of green-black firs. In our islands the Osprey is too rare to be gregarious or social; but in North America, where it is a very abundant species, great numbers of birds frequently live in colonies. The Osprey appears to pair for life, and yearly to return to its old breeding-place, some of its eyries having been occupied time out of mind. At the present day, in the Highlands, the Osprey usually makes its nest on the flat top of a pine tree, but formerly it just as frequently selected a battlement or chimney of some ruin, generally on an island. The nest is an immense pile of sticks, as much as four feet high and as many broad—the accumulation of many years—intermixed with turf and other vegetable matter, lined with finer twigs and finally with grass, much of it often green. The cavity containing the eggs is about twelve inches across and somewhat shallow. The Osprey is a light sitter, usually leaving the nest long before it is very closely approached.

RANGE OF EGG COLOURATION AND MEASUREMENT : The eggs of the Osprey are usually two or three in number, sometimes as many as four, and remarkably handsome. They vary in ground colour from white to pale buff, very boldly and richly blotched, and spotted with deep reddish-brown, and with underlying markings of violet-gray. Some varieties are so heavily marked that all trace of ground colour is concealed on the larger end of the egg; other varieties are suffused with a purple or reddish-orange shade; others have bold irregular blotches here and there over the entire surface, or principally distributed in a zone round either end, or round the middle; whilst others, yet again, are speckled and spotted with pale orange-red and violet-gray. Some are handsomely blotched with chestnut-red and blurred with gray. Average measurement, 2·3 inches in length, by 1·85 inch in breadth. Incubation, performed chiefly by the female, lasts from three to four weeks.

DIAGNOSTIC CHARACTERS: The size, combined with the exceedingly handsome appearance, readily distinguish the eggs of the Osprey from those of all other species breeding in the British Islands.

Family PHALACROCORACIDÆ.   Genus PHALACROCORAX.

CORMORANT.

PHALACROCORAX CARBO (*Linnæus*).

Single Brooded.   Laying season, April and May.

BRITISH BREEDING AREA: The Cormorant is a common and very widely dispersed species, breeding in more or less abundance on all the rocky coasts of the British Islands, including St. Kilda, the Orkneys, and the Shetlands. It also breeds at several inland stations, especially in Ireland and Wales.

BREEDING HABITS: The Cormorant is a resident in our islands, but is much more generally dispersed during autumn and winter. Its favourite breeding-haunts are ranges of lofty sea-cliffs, and small islands and reefs; but many birds frequent inland sheets of water, and breed near them, miles from the sea. The Cormorant is a gregarious bird, especially during the nesting season, and many of its colonies are large. It is not improbable that this species pairs for life, as the same breeding-places, in many cases the same nests, are tenanted annually. Such birds as have not paired, however, do so early in the spring, and all begin to congregate at the various nesting-stations. The nest of the Cormorant may be either on the ground on a low island or reef, on the ledges of maritime or other precipices, or in lofty trees and bushes. When on a low reef it is usually a mass of stalks of the coarser marine herbage, and sea-weed, lined with bits of green thrift, sea-campion, and sea-parsley; when on trees and rock ledges it is generally a great pile of sticks and twigs, lined with coarse grass, but the practice of adding green vegetation of some kind is still adhered to. The birds are not very demonstrative at the nests, leaving them before their colony is very closely approached, and usually retiring to some distance to await the passing away of the disturbance. I have frequently noticed that when the nests have been built on cliffs the birds are more loth to leave them than when placed on the low surface of an island.

RANGE OF EGG COLOURATION AND MEASUREMENT: The eggs of the Cormorant are from three to six in number, but the former amount is most frequently found. The colour of the shell is a delicate green, but usually this is entirely concealed (or only visible here and there) by a thick coating of lime, which may be easily removed with a pen-knife. The eggs are long and oval, and vary a good deal in size. Average measurement, 2·7 inches in length, by 1·6 inch in breadth. Incubation, performed by both sexes, lasts a month.

DIAGNOSTIC CHARACTERS: The eggs of the Cormorant may generally be distinguished from those of the Shag (the only species with which they can be confused in our islands) by their larger size; but this character is not constant, so that they require careful identification. It might be remarked that the Shag nests very frequently in a cave or fissure, or even in a hole in the cliffs.

Family PHALACROCORACIDÆ. Genus PHALACROCORAX.

SHAG.

PHALACROCORAX GRACULUS (*Linnæus*).

Single Brooded. Laying season, May and beginning of June.

BRITISH BREEDING AREA: The Shag is widely and generally dispersed along our entire coast, wherever such is rocky enough, and contains caves and fissures in which the bird may find suitable breeding-places. These are least frequent on our eastern coast-line, and the Shag consequently is there more local and less abundant than in more precipitous areas elsewhere.

BREEDING HABITS: The Shag is a resident in our islands, but given to much local movement during autumn and winter. Its favourite nesting-haunts are ranges of maritime cliffs which contain caves and fissures, but where such are not available the bird often contents itself with a rocky island, and breeds amongst the boulders near the beach. The Shag is a gregarious bird, but owing to its partiality for a cave or a fissure, the colonies vary considerably in extent, owing to the amount of accommodation afforded. In some cases the community consists of only a few pairs; in others of considerable numbers; whilst not unfrequently odd pairs may be met with in spots where there is no room for more. I am of opinion that the Shag, like the Cormorant, pairs for life, yearly frequenting the same nesting-place. The nests are either wedged into some crevice in the sides or roof of, or built on ledges in, an ocean cave; or in holes in the face of the cliff, some of them so small at the entrance as scarcely to admit the parent birds. Less frequently they are made amongst the strewn rocks and huge boulders near the beach of an islet or reef, or on a ledge of the cliffs, usually where the rocks overhang considerably. The nests are bulky structures, wherever the sites admit of elaborate building, composed externally of sticks, stalks of plants, and sea-weed, and lined with straw, coarse grass, and turf, all more or less matted together with droppings, decaying fish, and slime. Many nests are enlarged and patched up, season by season. The birds sit more closely than Cormorants usually do, many not leaving their nests until absolutely compelled. I have often had actually to drive this species from its eggs.

RANGE OF EGG COLOURATION AND MEASUREMENT: The eggs of the Shag are usually three in number, sometimes four or even five, but more often two. They are elongated, and the shell is of a delicate green, where it is visible through the more or less thick coating of lime. Average measurement, 2·5 inches in length, by 1·5 inch in breadth. Incubation, performed by both sexes, lasts about a month. The eggs are usually sat upon as soon as laid.

DIAGNOSTIC CHARACTERS: As a rule the eggs of the Shag may be distinguished from those of the Cormorant by their smaller size, but the rule unfortunately is not absolute, consequently the eggs require careful identification. The situation of the nest is of some service in the matter of identifying them.

Family SULIDÆ. Genus SULA.

GANNET.

SULA BASSANA, *Brisson*.

Single Brooded.   Laying season, May and early June.

BRITISH BREEDING AREA: The Gannet is an abundant but very locally distributed bird in the British Islands. Its only breeding-places within their area are as follows:—England: Lundy Island (a few pairs); Wales: Grassholm, off the coast of Pembrokeshire (a small colony); Scotland (E. coast): the Bass Rock (a large colony); (N. coast), the Stack of Suleskerry (a large colony); (N.W. coast), Sulisker (the largest colony in our islands); (W. coast), Boreray and adjoining Stacks in the St. Kilda Group (a very large colony); Ailsa Craig (a large colony); Ireland (S.W. coast): Little Skellig (a large colony).

BREEDING HABITS: The Gannet is a resident in the British seas, but seldom comes near land for any lengthened period except to breed, and during autumn and winter wanders about considerably in quest of its finny prey. It is a very gregarious bird during the nesting season, congregating in large numbers at certain favoured spots year by year. The Gannets begin to return to their breeding-colonies early in spring, nest-building commencing towards the end of April. The rude nests are made on the ledges of the ocean cliffs, amongst the broken rocks at the summit of the precipices, or on the flat tops of pinnacles and stacks. Vast numbers of nests are made close together; in some places so much so that it is well-nigh impossible to walk amongst them without breaking the eggs. The nest is not very large, but generally so trodden out of shape as to resemble a mere heaped mass of material, caked and matted together with slime, droppings, and filth which smells most offensively, especially on a hot, close day. It is made of sea-weed, turf, straws, moss, stalks of thrift and campion, and is frequently patched up and increased whilst incubation is in progress. The cavity containing the egg is shallow, almost flat, but the nest itself is sometimes as much as a foot or more in height. As the birds nest in such close companionship, quarrels are of frequent occurrence. The Gannet is a close sitter, usually remaining on the nest until almost pushed off, uttering loud cries of angry remonstrance. It is a noisy bird at the breeding-place, and the din, loud enough at all times, becomes almost deafening when the colony is fairly aroused. Many birds may be seen watching your approach with suspicion, standing on their solitary egg and rolling it about from side to side preparatory to flight. No pen can do justice to the almost overwhelming scene of noisy confusion, as the birds in thousands skim and float about in the air, or stand and utter their loud, harsh cries upon their nests. For further details respecting the Gannet's colonies I must refer the reader to my work on *Our Rarer Birds*.

RANGE OF EGG COLOURATION AND MEASUREMENT: The egg of the Gannet, for, as previously remarked, but one is laid, is pale bluish-green, but generally so thickly coated and plastered with lime as to conceal all trace of the actual colour of the shell. It soon however becomes discoloured, through contact with the wet, dirty nest, and the feet of the parent bird, frequently so much so as to resemble the egg of a Kestrel in tint. Average measurement, 3·2 inches in length, by 2·0 inches in breadth. The Gannet will lay several eggs in succession if they are removed. Incubation, performed by both sexes, lasts about six weeks.

DIAGNOSTIC CHARACTERS: The large size and chalky coating readily distinguish the egg of the Gannet from those of all other species breeding in our islands with which it is at all likely to be confused.

Family ANATIDÆ.
Sub-family CYGNINÆ.

Genus CYGNUS.

MUTE SWAN.

CYGNUS OLOR (*Gmelin*).

Single Brooded. Laying season, March, April, and May.

BRITISH BREEDING AREA: There can be little doubt that the Mute Swan has never bred in our islands in a perfectly wild state. In a more or less domesticated condition, however, it breeds wherever man affords it his protection, consequently its distribution is a purely artificial one, and possesses no ornithological interest or value whatever.

BREEDING HABITS: The Mute Swan is of course a resident in our islands, but it is not improbable that occasionally a few really wild birds pay them a visit from the Continent during winter. In some cases this Swan is to a certain extent gregarious during the breeding season, numbers of birds nesting in somewhat close companionship, but I should say that generally each pair frequent some chosen spot from which all intruders are jealously driven. The Swan pairs for life, and yearly returns to one place to breed, generally making a new nest annually, but occasionally using the old one. The nest is a huge, conspicuous structure, usually placed on an island, or amongst the rank vegetation on the bank of the pool or stream. It is made of dead reeds, rushes, and dry grass, and lined with finer but similar materials, together with a little down and a few feathers. Both parents assist in its construction, as well as in adding to the structure from time to time during the progress of incubation, the male usually collecting, and the female arranging the materials. The male is also said to sit for a few days in the empty nest to shape and warm the interior, preparing it for the eggs. When not actually sitting the male Swan is generally near the nest, a watchful sentinel, ever ready to do battle in its defence. It should also be remarked that the eggs are invariably covered by the female before she leaves them, the male uncovering them before he takes his share of the task of incubation, and remaining on them until his mate has thoroughly dried her plumage after feeding.

RANGE OF EGG COLOURATION AND MEASUREMENT: The eggs of the Mute Swan vary from three to twelve, according to the age of the female. Young females commence to lay in their second year (sometimes not until their third or fourth), usually producing from three to five eggs; from seven to nine will be produced the next season, and at four years old, ten or twelve. They are greenish-white, or very pale green, rough in texture, and with little or no polish. Average measurement, 4·5 inches in length, by 3 inches in breadth. Incubation, performed by both sexes, lasts from five to six weeks, according to the state of the season.

DIAGNOSTIC CHARACTERS: The eggs of the Swan may be readily distinguished from those of any other species breeding in our islands by their large size.

Family ANATIDÆ. Genus ANSER.
Sub-family *ANSERINÆ*.

GRAY-LAG GOOSE.

ANSER CINEREUS, *Meyer*.

Single Brooded.   Laying season, April and May.

BRITISH BREEDING AREA: Nearly a hundred years have passed since the Gray-Lag Goose bred in the fens and marshes of East Anglia. Its principal breeding area now is in the Outer Hebrides, but the bird still continues to nest in Ross-shire, Sutherlandshire, and Caithness. In Ireland this Goose breeds in a semi-domesticated state at Castle Coole, in Co. Monaghan.

BREEDING HABITS: The Gray-Lag Goose is a resident in the British Islands, but its numbers are increased during winter by arrivals from more northern latitudes. The favourite breeding-grounds of this Goose are wild moors and swamps. I am of opinion that the bird pairs for life. It is perhaps more gregarious in winter than in summer, but even in the breeding season continues its social instincts, and numbers of pairs very often nest on a small area of ground. The nest is almost invariably made on the ground, either amongst the tall heather on the moors, or in the rank vegetation of the swamps, but occasionally it is said to be built on the ledge of a rock. It is a huge structure, as much as three feet in diameter at the base, and more than a foot high. Externally it is made, according to the locality, of branches and twigs of heather, dead rushes and reeds, dry grass, bracken, leaves, and turf, and lined with moss, and, as incubation progresses, more and more thickly with down and feathers, plucked by the female from her breast. This bird is rather a close sitter, and the male keeps constant watch and ward near the nest, ever ready to drive off intruding birds and beasts.

RANGE OF EGG COLOURATION AND MEASUREMENT: The eggs of the Gray-Lag Goose are from six to eight in number, but clutches of twelve and even fourteen have been recorded. They are creamy-white, and the shell is without polish. Average measurement, 3·45 inches in length, by 2·35 inches in breadth. Incubation, performed by the female, lasts about a month.

DIAGNOSTIC CHARACTERS: The present species is the only Goose breeding within our limits, consequently the eggs cannot readily be confused with those of any other British bird.

Family ANATIDÆ. Genus TADORNA.
Sub-family ANATINÆ.

COMMON SHELDRAKE.

TADORNA CORNUTA (*S. G. Gmelin*).

Single Brooded. Laying season, April and May.

BRITISH BREEDING AREA: The Common Sheldrake is widely if somewhat locally distributed throughout the coasts of the British Islands that are suited to its requirements. It breeds on the low sandy portions of the east and west coasts of England, but is much more local on the south coast. It also breeds on all the coasts of Scotland, including the Hebrides; and the same remarks apply to the low sandy coasts of Ireland, where, however, the bird is rarer and more local.

BREEDING HABITS: The Sheldrake is a resident on the British coasts, but subject to considerable local movement during the non-breeding season. Its favourite breeding-grounds, never very far from the sea,[1] are low-lying sandy coasts, dunes, links, flat sand-banks, and small islands in sea lochs, and estuaries. I am of opinion that this handsome Duck pairs for life, and in many, if not all, cases returns to one particular spot to breed. It is not a social or gregarious species in our islands, although numbers of pairs may be met with breeding along a small stretch of sandy coast. Each pair, however, keeps to itself. The nest is made at the extremity of a burrow, a rabbit-hole being frequently selected, but sometimes a hole among masonry or under rocks is chosen, whilst more rarely a spot is preferred in a dense gorse covert. The bird, exceptionally I consider, sometimes makes its own burrow, which is described as being in a nearly circular direction. The burrow is of various lengths, sometimes as much as twelve or fifteen feet, sometimes not more than half that distance. At the end, in a small chamber, the rude nest of dry grass is formed (a rabbit's nest is not unfrequently utilized), and as incubation advances this is warmly lined with down from the parent's body. Few nests are more difficult to find, the birds being remarkably cautious in leaving or visiting it. The locality may sometimes be indicated by the male flying round and round above the burrow, or the secret of the nest betrayed by the parents at morning and evening, when the sitting bird is relieved by its mate.

RANGE OF EGG COLOURATION AND MEASUREMENT: The eggs are from six to twelve in number, sometimes as many as sixteen; and in cases where they have been removed as many as thirty have been taken in a season from a single burrow. They are creamy-white, very

[1] Instances of this species breeding on the heaths of Dersingham and Sandringham, and on a farm at Sedgeford, thirty years ago, are recorded in Stevenson's *Birds of Norfolk*.

brittle, smooth, and with considerable polish. Average measurement, 2·7 inches in length, by 1·9 inch in breadth. Incubation, performed by both sexes, the female sitting the most, lasts about a month.

DIAGNOSTIC CHARACTERS: The eggs of this and most other species of Ducks are readily and safely identified by the down of the parent bird, which forms a lining to the nest. It is most important, therefore, that the collector should take at least a portion of the down and a few of the small feathers from every nest, a proceeding which will tend to authenticate his specimens, and immensely increase the interest and value of his collection. The down of the Sheldrake is lavender-gray, mixed with a few almost white tufts. Armed with the down, the eggs of the present species cannot be confused by the collector with those of any other Duck breeding in our islands.

Family ANATIDÆ.   Genus ANAS.
Sub-family ANATINÆ.

GADWALL.

ANAS STREPERA, *Linnæus.*

Single Brooded. Laying season, May.

BRITISH BREEDING AREA: The Gadwall is only known to breed in one portion of our islands, in the county of Norfolk. Originally only a pair of pinioned birds that had been taken in Dersingham Decoy, and turned loose on the lake at Narford, upwards of forty years ago, were known to breed; but since that time their descendants have continued to do so, as well as to attract really wild birds, so that at the present day this Duck breeds in some considerable numbers, not only at Narford, but at Merton and some few other neighbouring localities.

BREEDING HABITS: The Gadwalls that breed in England are probably resident, but their numbers are increased during the winter by arrivals from the Continent. The Gadwall frequents fresh water rather than the coast, its favourite breeding-grounds being marshy heaths and the boggy banks of lakes and meres. Although several nests may be met with on a comparatively small area of ground, the Gadwall is neither gregarious nor social during the breeding season, each pair keeping to themselves. It is not improbable that this species pairs for life, and yearly frequents a chosen place in which to breed. The nest is rarely if ever placed far from water, and is frequently made on the shore of a low island or even on a clump of herbage surrounded by water. The favourite site seems to be a tussock of sedge, or beneath the shelter of a tuft of coarse herbage or reeds in a bog. The nest, which is deep and well put together, is made of dead leaves, or dry grass and sedge, somewhat sparingly lined with down and a few feathers. The bird is rather a close sitter, but makes little demonstration at the nest, although when disturbed with her brood she will feign lameness, just as most other species of this family do.

RANGE OF EGG COLOURATION AND MEASUREMENT: The eggs of the Gadwall are from eight to thirteen in number, ten being an average clutch. They are creamy-yellow, smooth in texture, and show considerable polish. Occasionally a scarcely perceptible green tinge is perceptible. Average measurement, 2·1 inches in length, by 1·5 inch in breadth. Incubation, performed by the female, is said by some authorities to last about twenty-five days; but according to Naumann only twenty-one or twenty-two days.

DIAGNOSTIC CHARACTERS: The only safe guide in identifying the eggs of this species is by the down, which is brownish-gray, the tufts smaller than those of the down of the Mallard, and the pale tips almost imperceptible.

Family ANATIDÆ.
Sub-family *ANATINÆ.*
Genus ANAS.

PINTAIL DUCK.

ANAS ACUTA, *Linnæus.*

Single Brooded.   Laying season, May.

BRITISH BREEDING AREA: The Pintail Duck is another exceedingly rare and local species in our islands during summer. It has been found breeding on Hysgeir, off the south coast of Skye; and I have every reason to believe that it does so on certain small rocky islets in the Firth of Forth. In Ireland a few pairs breed at Abbeyleix in Queen's County, on Loughs Mask and Corrib in County Galway, and probably in some parts of Connemara.

BREEDING HABITS: It is difficult to say whether the Pintails that breed in our islands are resident, as in autumn great numbers of this Duck visit them from the north, and remain until the following spring. The breeding-haunts of this species in our islands are rocky inlets in quiet loughs and firths, or at some considerable distance from the mainland. In other countries, swamps and moors, and the margins of lakes and ponds are described as its favourite retreats during the nesting season. The birds appear to pair annually, but I suspect many remain mated for life. The nest is made upon the ground amongst coarse grass and other vegetation, or under shrubs, or beneath the shelter of a rock. It is composed of dry grass, withered sedges and rushes, and dead leaves, lined as the eggs are laid and incubation progresses with an abundance of down from the body of the female. The bird sits closely, but makes little or no demonstration when flushed.

RANGE OF EGG COLOURATION AND MEASUREMENT: The eggs of the Pintail are from six to ten in number, pale buffish-green in colour, smooth in texture, but with little gloss. Average measurement, 2·15 inches in length, by 1·5 inch in breadth. Incubation, performed by the female, lasts from twenty-three to twenty-seven days.

DIAGNOSTIC CHARACTERS: The eggs of the Pintail cannot always be distinguished from those of the Mallard, but the down from the nest is a tolerably safe guide to their identification, being sooty-brown in colour, distinctly tipped with white, but not so conspicuously as that of the Wigeon.

Family ANATIDÆ.   Genus ANAS.
Sub-family *ANATINÆ*.

WIGEON.

ANAS PENELOPE, *Linnæus*.

Single Brooded. Laying season, May.

BRITISH BREEDING AREA: The Wigeon, of course, is best known as a winter migrant to the British Islands, but a few breed in Ross-shire, Sutherland, Caithness,

Cromarty, the Orkneys and Shetlands. In Ireland it is said to breed locally and sparingly in Counties Antrim, Armagh, Tyrone, and Mayo.

BREEDING HABITS: It is difficult to say whether the Wigeons that breed with us retire south in autumn or not, their personality being utterly effaced by the individuals that pour in from the north at that season. The breeding-haunts of this Duck are the rough districts on the borders of the moors close to the limit of forest growth—scrubby woodlands, swamps, and heaths clothed with a coarse vegetation, and studded with lakes and tarns and streams. We still remain in ignorance as to whether this species pairs for life or not: personally I should incline to the latter view, after what I have observed of its economy. So far as our islands are concerned the Wigeon cannot be regarded as at all gregarious during the summer, but in many localities the nests are sprinkled pretty closely over suitable ground, suggesting at least a social tendency. The nest is made in a variety of situations, sometimes at a considerable distance from water, but more usually near the lake or tarn-side. Sometimes it is made amongst heather, at other times in coarse grass, or beneath the shelter of a stunted bush. The nest is made of dry grass and withered fragments of aquatic herbage, warmly lined with down from the body of the female. The bird is a close sitter, but not demonstrative when flushed from the eggs.

RANGE OF EGG COLOURATION AND MEASUREMENT: The eggs of the Wigeon are from six to ten in number; in rare instances as many as twelve. They range from creamy-white to buffish-white in colour, are smooth in texture, but with little polish. Average measurement, 2·2 inches in length, by 1·5 in breadth. Incubation, performed by the female, lasts from twenty-four to twenty-five days.

DIAGNOSTIC CHARACTERS: The eggs of the Wigeon never show any trace of *olive* in their colouration, so cannot easily be confused with those of other species. They somewhat closely resemble those of the Gadwall, but are a trifle larger; the down, however, serves to distinguish them, and is sooty-brown with distinct white tips.

Family ANATIDÆ.　　　　　　　　　　　　Genus ANAS.
Sub-family *ANATINÆ*.

COMMON TEAL.

ANAS CRECCA, *Linnæus*.

Single Brooded.　Laying season, May.

BRITISH BREEDING AREA: The Common Teal is locally distributed throughout the British Islands, but becomes rarer in the Hebrides, the Orkneys, and the Shetlands. So far as the mainland is concerned this Duck is certainly commoner in the eastern and northern counties of England than in the southern and western counties, and is even more so in Scotland. In Ireland it is also locally but fairly well dispersed.

BREEDING HABITS: The individual Teals that breed with us may or may not be resident in our islands; and the question is somewhat difficult to decide, for great numbers of this species visit us in autumn from more northern localities. The nesting haunts of this charming little species are the margins of ponds and small lakes, especially such that are situated in swamps or marshy country, and where the shore is fringed with a good growth of reeds, flags, iris, and other coarse vegetation. I am of opinion that this Duck pairs for life (a fact which will doubtless be found to apply to the Anatidæ in general). The nest is made upon the ground, under brambles or amongst heath, sedge, and coarse grass, either growing by the water-side, or less frequently at some considerable distance from the pool in a wooded swamp. I always think a Teal's nest is a charming little structure, made as it is of dry grass, bits of fern-frond, broken sedges and reeds, and lined warmly with the down from the female's body. The female is a close sitter, but not demonstrative when flushed from the nest.

RANGE OF EGG COLOURATION AND MEASUREMENT: The eggs of the Common Teal are from eight to ten in number; in rare cases as many as fifteen. They vary from creamy-white to buffish-white in colour, sometimes with a faint greenish tinge. Average measurement, 1·7 inch in length, by 1·3 inch in breadth. Incubation, performed by the female, lasts from twenty-one to twenty-two days.

DIAGNOSTIC CHARACTERS: The eggs of the Teal cannot be distinguished with certainty from those of the Garganey, but fortunately the down in the nest is a sure guide to their identity, the tufts being small and uniform dark-brown without any pale tips. From the eggs of all other Ducks breeding in our area the eggs of the present species may be at once distinguished by their small size.

Family ANATIDÆ. Genus ANAS.
Sub-family ANATINÆ.

GARGANEY.

ANAS CIRCIA, *Linnæus.*

Single Brooded. Laying season, May.

BRITISH BREEDING AREA: The Garganey is yet another of our rarest and most local birds, and is perhaps only known with certainty to breed in the Broad districts of Norfolk and Suffolk. It may possibly do so in some of the more southern English counties, but the information is as yet meagre and indefinite.

BREEDING HABITS: The Garganey is a summer migrant to the British Islands, reaching its breeding-grounds in East Anglia towards the end of February and in March. Its nesting haunts are the rough marshy lands adjoining the open broads and pools—more or less reclaimed areas of ground studded with tufts of rushes and hummocks of sedge. The Garganey probably pairs for life, and appears to migrate in pairs and to swim in company until the nesting season. The nest of this species is made in a great variety of situations, usually on the ground, but an instance is on record where it was discovered in the stump of a willow tree. Frequently it is placed near a footpath or even the public highway. The favourite site appears always to be the centre of a tuft of sedge, coarse grass, or rushes; occasionally it is made in long grass, heather, or even in growing corn. It is a rather deep structure, made of dry grass, dead rushes, leaves, and other vegetable *débris*, warmly lined with down. The female is a remarkably close sitter, but when flushed makes little or no demonstration.

RANGE OF EGG COLOURATION AND MEASUREMENT: The eggs of the Garganey are from eight to fourteen in number. They vary from creamy-white to buffish-white in colour, are smooth in texture, but with little polish. Average measurement, 1·8 inch in length, by 1·35 inch in breadth. Incubation, performed by the female, lasts from twenty-one to twenty-two days.

DIAGNOSTIC CHARACTERS: The eggs of the Garganey cannot be distinguished from those of the Teal, the only species with which they can possibly be confused; the down in the nest, however, readily settles their identity, the tufts being small and brown, *with long white tips.*

Family ANATIDÆ. Genus ANAS.
Sub-family *ANATINÆ.*

SHOVELLER.

ANAS CLYPEATA, *Linnæus.*

Single Brooded. Laying season, May.

BRITISH BREEDING AREA: In England the Shoveller is known to breed more or less sparingly in the counties of Dorset, Kent, Hertford, Cambridge, Norfolk, Lincoln, Notts, Huntingdon, Stafford, York, Durham, Northumberland, and Cumberland. In Scotland, Kircudbright, East Lothian, Dumbarton, Argyle, Elgin, Ross, Sutherland, and the island of Tiree in the Hebrides. In Ireland, in Queen's County, Galway, Dublin, and Antrim.

BREEDING HABITS: It is doubtful whether the individuals of this species that breed in our islands are resident in them; the evidence seems to suggest a southern movement in winter. The nesting haunts of the Shoveller are the rough marshy lands and swampy heaths in the vicinity of broads, meres, and sluggish, weed-choked streams. Plantations of small trees and belts of wooded country are also resorted to if favourably situated near to waters frequented by this species. The Shoveller, if not exactly gregarious during the breeding season, is certainly social, and several nests may be found within a comparatively small area of suitable ground. The nest is made in a variety of situations, in a tuft of sedge, amongst coarse long grass on a bank, or growing grain, or on dryer ground in heath. It is a mere hollow, scantily lined with a little dry grass, sedge, or dead leaves, and lined with a fair amount of down and small feathers from the body of the female. The bird is a close sitter, and when flushed from the eggs makes little or no demonstration; the case is very different, however, when the young are hatched.

RANGE OF EGG COLOURATION AND MEASUREMENT: The eggs of the Shoveller are from seven to fourteen in number, nine or ten being an average clutch. They are pale buffish-white with a faint tinge of olive-green (more pronounced on some eggs than others), fine in texture, and with some little polish. Average measurement, 2·0 inches in length, by 1·5 inch in breadth. Incubation, performed almost invariably by the female (in one case at least the cock has been flushed from the eggs), lasts according to Naumann from twenty-one to twenty-three days.

DIAGNOSTIC CHARACTERS: The eggs of the Shoveller require careful identification, for those of the Pintail and the Mallard come very close in general appearance. The down, however, is pretty characteristic, and should prevent confusion, the tufts being of moderate size, neutral dark gray with pale centres and very conspicuous white tips.

Family ANATIDÆ.
Sub-family ANATINÆ.
Genus ANAS.

MALLARD.

ANAS BOSCHAS, *Linnæus*.

Single Brooded. Laying season, March, April, May (exceptionally even in February).

BRITISH BREEDING AREA: The Mallard is commonly and generally distributed throughout the British Islands in all districts suited to its requirements, extending to the Hebrides, the Orkneys, and Shetlands. It is not so abundant as a breeding species in England, perhaps, as was formerly the case, before drainage destroyed many of its haunts, but it still continues fairly common.

BREEDING HABITS: The Mallard is a resident in the British Islands, but largely increased in numbers during autumn and winter by arrivals from other lands. The nesting haunts of this ubiquitous species are of a very varied character, and extend from the marshes and swamps of the lowlands to the mountain moors and wet upland wastes. Water is by no means essential, and yet in some districts there is scarcely a pond or a stream that is not frequented by a pair of birds. The bird is not perhaps gregarious during the breeding season, but it is to a very great extent a social one, and several nests may often be found within a few yards. The Mallard unquestionably pairs for life. The nest is made in a great variety of situations, and by no means always on the ground. I have taken nests of this Duck in open parts of Sherwood Forest, on ground covered with bracken and studded with thorn trees; also on the barest ground under long heather on small islands in the Highland lochs. Sometimes a deserted nest of a Crow or a Rook, a Wood Pigeon or a Hawk is utilized. Occasionally it is made under the shelter of a peat-wall, in a boat-house, in a hollow tree-trunk, on the top of a pollard; more frequently in a field of corn, or a hedge-bottom. Very often it is made amongst long rank grass by the water-side, or in a tuft of rushes or sedge. It is usually made in a hollow scraped in the ground, and is composed of dry grass, bracken leaves, moss, heath, or whatever vegetable refuse is to be obtained in the vicinity, and warmly lined with down and a few small feathers from the body of the female. Some nests of this species are remarkably handsome, and stand as much as eight or ten inches above the level of the ground. The bird is a close sitter, and when flushed makes little or no demonstration, hiding herself as soon as possible in the nearest cover.

RANGE OF EGG COLOURATION AND MEASUREMENT: The eggs of the Mallard are from eight to sixteen in number, twelve being an average clutch. They vary in colour from pale buffish-green to greenish-buff, are fine and smooth in texture, and with a faint polish. Average measurement, 2·3 inches in length, by 1·6 inch in breadth. Incubation, performed by the female, lasts from twenty-six to twenty-eight days. The eggs of this species, as is almost universally the case in the Anatidæ, are covered for concealment when the parent voluntarily leaves the nest.

DIAGNOSTIC CHARACTERS: The eggs of the Mallard rather closely resemble those of the Pintail and the Shoveller, but the down in the nest renders their identification safe. The tufts of this are large, neutral gray in colour, with very faint white tips.

Family ANATIDÆ. Genus FULIGULA.
Sub-family FULIGULINÆ.

POCHARD.

FULIGULA FERINA (*Linnæus*).

Single Brooded. Laying Season, May.

BRITISH BREEDING AREA: The Pochard breeds in Lancashire, the East Riding of Yorkshire, some of the Midland Counties, Dorset, and perhaps most abundantly of all in Norfolk. In Scotland it is known to breed in South Perthshire and in Fifeshire; whilst in Ireland it does so in Cos. Sligo, Antrim, and Tipperary.

BREEDING HABITS: It is difficult to say whether the Pochards that breed with us draw south in winter, as in autumn large numbers of this species visit our islands and intermingle with them. The breeding haunts of this species are large open sheets of water where plenty of cover in the shape of reeds, rushes, sedges, iris, and the like clothe the margin, or where the pools are surrounded by slightly higher ground covered with heath, furze, and tufts of rushes and coarse grass. The Pochard is certainly a most social species during the breeding season, and several nests may often be found close together. The nest of this Duck is always made near fresh water, and in many instances is a floating structure built on a mass of fallen vegetation several yards from shore, or in a tussock surrounded by shallow water. A favourite situation is in a tussock of vegetation of some kind, notably *carex*, less frequently *Scirpus lacustris*. Sometimes it is made amongst a bed of flags or iris, or in a crown of rushes. The nest is made of dry grass, sedge, broken rushes and flags, or any similar aquatic vegetation readily obtainable, and lined with down and a few feathers from the body of the female. The bird is a closer sitter, but when flushed makes little or no demonstration.

RANGE OF EGG COLOURATION AND MEASUREMENT: The eggs of the Pochard are from eight to twelve or even fourteen in number, ten being an average clutch. They are brownish-gray or greenish-drab in colour, and smooth in texture. Average measurement, 2·4 inches in length, by 1·7 inch in breadth. Incubation, performed by the female, lasts from twenty-five to twenty-eight days.

DIAGNOSTIC CHARACTERS: The eggs of the Pochard cannot be distinguished from exceptionally large eggs of the Tufted Duck, but the down may be taken as an important aid to their correct identification. In the present species the down tufts are large, grayish-brown in colour, with dull white centres.

Family ANATIDÆ. Genus FULIGULA.
Sub-family *FULIGULINÆ*.

TUFTED DUCK.

FULIGULA CRISTATA, *Leach*.

Single Brooded.   Laying season, latter half of April, May, and June.

BRITISH BREEDING AREA: The Tufted Duck breeds locally throughout the British Islands. In England it is known to do so in Northumberland, Lancashire, Yorkshire, Shropshire, Notts, Norfolk, Sussex, Hertfordshire, Dorset, and Devonshire (at Slapton Ley). In Scotland it breeds in Roxburghshire, Perthshire, Kinross-shire, and Aberdeenshire ; and in Ireland near Loughs Neagh and Beg, and in some parts of Co. Monaghan.

BREEDING HABITS: The Tufted Duck is by far the most abundant in winter, but it is hard to say whether the birds that breed in our area retire south as the northern contingents arrive. In its choice of a breeding-haunt the Tufted Duck closely resembles the Pochard, showing partiality for pools and meres and broads surrounded with rough, hummocky, marshy land, or heathy and fairly well timbered ground. This Duck probably pairs for life. I have seen it in pairs at every season of the year. It is also a remarkably social species during the breeding season, the males not only swimming in company, but the females making their nests at no great distances apart in many cases. The favourite situation for the nest is a tuft or tussock of sedge, amongst rushes, in long coarse grass, or beneath the shelter of a stunted bush. Sometimes a heap of dead reeds is selected. The nest is a mere hollow, lined with a little dry grass, sedge, or rush, and an abundance of down from the body of the female, not unfrequently intermixed with a few feathers from the male. The bird is a close sitter, but when flushed flies straight away without any alluring or anxious movement.

RANGE OF EGG COLOURATION AND MEASUREMENT: The eggs of the Tufted Duck are usually from eight to ten in number, but sometimes as many as thirteen or even fourteen. They are greenish-buff in colour, smooth in texture, and rather polished. Average measurement, 2·3 inches in length, by 1·6 inch in breadth. Incubation, performed by the female, lasts from twenty-five to twenty-eight days. The eggs are always covered by the female for concealment when she leaves them voluntarily.

DIAGNOSTIC CHARACTERS: The eggs of the Tufted Duck cannot safely be distinguished from those of the

Pochard, but the down in the nest serves to identify them. The tufts are small and grayish-black in colour, with obscure pale centres—much darker than that of the preceding species.

Family ANATIDÆ. Genus FULIGULA.
Sub-family *FULIGULINÆ.*

COMMON SCOTER.

FULIGULA NIGRA (*Linnæus*).

Single Brooded. Laying season, May.

BRITISH BREEDING AREA: The Common Scoter only deserves its name during autumn and winter, when its numbers often blacken the seas, but during the breeding season it is a rare and excessively local bird. It is at present only known to breed in small numbers in Caithness, Sutherlandshire, and Ross-shire. It is also said to breed on the Earnly Marshes near Chichester.

BREEDING HABITS: The Common Scoters that breed with us so utterly lose their identity in the countless hordes that pour south in autumn and winter, that it is a matter of impossibility to say whether they move more to the south of us at that season or not. The favourite breeding-haunts of this Duck are the moorland lakes and rivers close to the sea, especially in such localities where small birches and willows abound amongst the broken, heath-clothed ground. I do not trace much sociability in this species during the breeding season, but unfortunately naturalists who have studied the habits of the bird in regions where it is abundant fail to inform us of the matter. Probably this species pairs for life, although the data on which the statement is founded is more of a general than of a special character. The nest, placed on an island, if such is to be had, is merely a hollow lined with a little dry grass, sprigs of heather, dead leaves, and similar refuse, and finished off with a bed of down from the body of the female. The bird sits closely, but when flushed flies right away without manifesting concern for the safety of the eggs.

RANGE OF EGG COLOURATION AND MEASUREMENT: The eggs of the Common Scoter are from six to nine in number, and pale grayish-buff or yellowish-white in colour, smooth in texture, and with little polish. Average measurement, 2·5 inches in length, by 1·8 inch in breadth. Incubation, performed by the female, is said to last about twenty-eight days. The eggs are covered for concealment when left voluntarily by the female.

DIAGNOSTIC CHARACTERS: The eggs of the Common Scoter cannot readily be confused with those of any other species breeding in our area, their size and colour combined distinguishing them. The eggs of the Goosander resemble them in colour and size, but are *heavier*, whilst the colour of the down prevents any confusion. The down tufts of the Scoter are large, brownish-gray in colour, with pale centres; that of the Goosander is grayish-white.

Until more reliable evidence is forthcoming, I must decline to admit the Golden-Eye (*Clangula glaucion*) into a work which deals exclusively with species that breed within the confines of the British Archipelago. The bright green colour of the eggs, and the nest placed in a hollow tree, will readily serve to distinguish them whenever they may be fortunately discovered.

Family ANATIDÆ. Genus SOMATERIA.
Sub-family *FULIGULINÆ*.

COMMON EIDER.

SOMATERIA MOLLISSIMA (*Linnæus*).

Single Brooded. Laying season, May and June; July occasionally.

BRITISH BREEDING AREA: The Eider Duck is a decidedly northern species, and is only known to breed in one locality in England, viz. on the Farne Islands. Northwards it becomes more generally, if locally, distributed, and may be found breeding from the Firth of Forth onwards in all suitable localities round the Scotch coast as far south as Inverness, including the adjoining islands, such as the Orkneys and Shetlands, the Hebrides, and St. Kilda. It is not known to breed anywhere on the Irish coast-line.

BREEDING HABITS: The Eider is a thoroughly maritime species, and only very exceptionally nests at any considerable distance from the sea. Its favourite breeding-haunts are rocky islands, low in elevation, and well covered with marine herbage. It is probable that this species pairs for life. Late in spring the flocks that have lived in company during the winter begin to separate more distinctly into pairs, and at this season there is often considerable rivalry displayed amongst the males. The female alone makes the nest, the male rarely if ever visiting it; but he is usually to be met with at sea close by the islands where his mate is brooding. The nest is invariably placed on the ground, usually amongst bladder-campion or long coarse herbage, often on a ledge of rocks or in a crevice. I have seen it on the edge of the cliffs, several hundreds of feet above the sea. It is a bulky, well-made structure, composed of coarse grass, dry sea-weed, heather, and bits of dead vegetation, lined profusely with down and a few curly feathers from the body of the female, gradually accumulated as the eggs are laid. Numbers of nests may be found close together, the birds being more or less gregarious throughout the year. The bird sits remarkably close, only leaving the eggs when absolutely compelled, and often allowing herself to be stroked by the hand, especially in districts where the birds are protected for their highly-prized down.

RANGE OF EGG COLOURATION AND MEASUREMENT: The eggs of the Common Eider are from five to seven in number, exceptionally as many as eight. They range in colour from creamy-gray to grayish-green or olive-green, and are smooth and wax-like in texture. Average measurement, 3 inches in length, by 2 inches in breadth. Incubation, performed by the female, lasts twenty-eight days.

DIAGNOSTIC CHARACTERS: The large size of the eggs of the Common Eider readily distinguish them from those of all other species breeding in our area. Down tufts, moderate in size, and varying from brownish-gray to grayish-brown, with obscure pale centres.

Family ANATIDÆ.   Genus MERGUS.
Sub-family *MERGINÆ*.

GOOSANDER.

MERGUS MERGANSER, *Linnæus*.

Single Brooded. Laying season, April and May.

BRITISH BREEDING AREA: The Goosander is another extremely local species, and only breeds in a few localities in the Highlands. Up to the present time it is certainly known to breed in Sutherlandshire, Argyleshire, and Perthshire. It seems probable that the bird is on the increase as a breeding species in our area, and may yet be detected doing so in Ireland.

BREEDING HABITS: The favourite breeding-haunts of the Goosander are open swampy forests full of lakes and rockbound streams. It is not improbable that this species pairs for life, and yearly resorts to one locality to breed. The nest is generally placed in a hole in a tree, but in localities where such is not available, a cleft or crevice in a rock or cliff, or a cavity amongst exposed tree-roots by the water-side is used instead. The nest is slight, especially when in a hole in a tree, when the dust at the bottom serves for the bed of the first eggs, but as the full clutch is laid, a warm lining of down is added. The bird is a close sitter.

RANGE OF EGG COLOURATION AND MEASUREMENT: The eggs of the Goosander are from eight to twelve in number; sometimes thirteen have been found. They are creamy-white in colour, glossy, and smooth in texture. Average measurement, 2·7 inches in length, by 1·8 inch in breadth. Incubation, presumably performed entirely by the female, lasts twenty-eight days.

DIAGNOSTIC CHARACTERS: The situation of the nest, the colour of the down (tufts large, and uniform grayish-white), and the creamy tint of the eggs prevent any confusion with those of other allied species breeding in our area.

Family ANATIDÆ. Genus MERGUS.
Sub-family MERGINÆ.

RED-BREASTED MERGANSER.

MERGUS SERRATOR, *Linnæus*.

Single Brooded. Laying season, May and June.

BRITISH BREEDING AREA: It is somewhat remarkable that the Red-breasted Merganser does not breed in England, seeing that it does so in the same latitude in Ireland. From the Clyde northwards the present species breeds in all suitable localities, both inland and maritime, up to the Orkneys and the Shetlands, and west to the Hebrides, but not, so far as I can learn, to St. Kilda. In Ireland it is equally widely dispersed, both inland and on the coasts, but appears to be much less numerous.

BREEDING HABITS: The principal breeding-grounds of the Red-breasted Merganser are the quiet, secluded shores of lochs and inland waters, and low rocky islands, especially such as stud the fjords and inlets of the coast. This species probably pairs for life. It can scarcely be regarded as gregarious during the breeding season, although numbers of nests may be found within a small area of suitable ground. The nest is usually made under the shelter of a rock or a bank, but rabbit-burrows and crevices in walls are sometimes selected. Occasionally it is made amongst long heath or furze close to the water-side. The nest is scanty enough; in many cases dispensed with altogether, the eggs lying on the bare ground until sufficient down accumulates to cover them. It is merely a slight arrangement of dry grass and leaves, but eventually the warm lining of down is added which makes a luxurious bed for the eggs. The hen is a close sitter, and when disturbed slips quietly from the eggs and quits the place with little or no demonstration. The male is never seen at the nest, but is usually to be met with on the water adjoining.

RANGE OF EGG COLOURATION AND MEASUREMENT: The eggs of the Red-breasted Merganser are from eight to twelve in number. They are uniform olive-gray, of various shades in colour, smooth in texture, and somewhat glossy. Average measurement, 2·6 inches in length, by 1·7 inch in breadth. Incubation, performed by the female, lasts twenty-eight days.

DIAGNOSTIC CHARACTERS: The eggs of the present species may be confused with those of the Scaup—those of the Pochard are much smaller—but the down (tufts large, pale brownish-gray, with obscure pale centres and tips) readily prevents confusion, and should be taken in every case for correct and perfect identification.

Family ARDEIDÆ. Genus ARDEA.

COMMON HERON.

ARDEA CINEREA, *Linnæus*.

Single Brooded. Laying season, March and April.

BRITISH BREEDING AREA: The Common Heron is widely distributed throughout the British Islands, breeding in every part, with the exception of the treeless Outer Hebrides, the Orkneys, and the Shetlands. Its colonies, however, are local, and in Scotland and Ireland small in comparison with those in England.

BREEDING HABITS: The Heron is a resident in our islands. Its favourite breeding-places are woods, plantations, and groves of trees on islands, but in localities where suitable trees cannot be found, a ledge of a cliff, a ruin, or even the ground is chosen. The presence of water is not essential to a heronry. This bird breeds in societies like Rooks, and as it probably pairs for life, yearly returns to one favourite spot to breed. Some of our British heronries have been in use from time immemorial. The nest of this species is built in a great variety of situations, on trees of all kinds, especially firs and larches, on ivy-clad ruins, on the ledges of crags and cliffs, and amongst heather on the hill-sides. In some cases a heronry will be established in or near a rookery. The nest is usually a bulky flat platform of sticks, generally at some distance from the trunk on a broad horizontal branch, less frequently on the top of a tree, or in a wide fork close to the stem. The finer sticks are used for the interior, which is sometimes further embellished with turf and moss. The nests vary a good deal in size, some of them being very large, and evidently the accumulation of years, and all are more or less whitewashed with droppings. Some trees contain but one nest, others two or three, according to the amount of accommodation offered, or the caprice of the birds. When the colony is invaded the big gray birds flutter from their nests, their wings crashing against the branches, and all is soon in silent commotion. As long as the intrusion lasts the birds continue to soar above their nests, now high, now low; every now and then a bird dropping on to its home as soon as the intruder is a safe distance away.

RANGE OF EGG COLOURATION AND MEASUREMENT: The eggs of the Heron are from three to five in number. They are greenish-blue in colour, more or less elliptical in shape, without polish, chalky, and rough in texture. They also vary a good deal in tint, some being much brighter and bluer than others. Average measurement,

2·5 inches in length, by 1·7 inch in breadth. Incubation, performed by both sexes, lasts twenty-five or twenty-six days.

DIAGNOSTIC CHARACTERS: The size, colour, and texture of the eggs of the Heron readily distinguish them from those of any other species breeding in our islands.

Before drainage and modern land improvement destroyed its strongholds, the Bittern (*Botaurus stellaris*) bred regularly in the fens and marshes of East Anglia, in various parts of Scotland, and in Ireland. There can be little doubt that this species does not breed anywhere in our islands at the present time. The fact could scarcely be overlooked, as the birds would be sure to attract attention by their singular booming cry. The eggs are laid in April and May, sometimes as early as March. The nest is built upon the swampy ground, amongst dense aquatic vegetation, and is composed of dead and rotting reeds, flags, and other herbage—a mere heap of rubbish, with a shallow cavity at the top. The eggs are from three to five in number, four being an average clutch. They are uniform brownish-olive or buff in colour, very similar in tint to those of the Pheasant. It may also be remarked that the interior of the shell when held up to the light is brown, not green as is universally the case with the true Herons. Average measurement, 2·1 inches in length, by 1·5 inch in breadth. Incubation, performed chiefly by the female, lasts from twenty-three to twenty-five days. Only one brood is reared in the year.

Family ŒDICNEMIDÆ. Genus ŒDICNEMUS.

STONE CURLEW.

ŒDICNEMUS CREPITANS, *Temminck.*

Single Brooded.   Laying season, May and June.

BRITISH BREEDING AREA: The Stone Curlew is another local species confined during the breeding season to the eastern and southern counties of England, as far north as Yorkshire, and as far west as Dorset. It breeds on the heaths and wolds of Yorkshire, Lincolnshire, Norfolk, Suffolk, and Cambridgeshire, southwards through Beds, Herts, Bucks, Oxfordshire, Berks, Wilts, Dorset, Hants, Sussex, and Kent. Exceptionally it has been known to nest in Worcestershire, Rutland, and Notts.

BREEDING HABITS: The Stone Curlew is a summer migrant to our islands, but a few individuals remain to winter in the extreme south-west of England, in Somerset, Devon, and Cornwall. It usually arrives in England in April. The breeding-haunts of this species are heaths, downs, sandy commons, and warrens—bare, treeless districts. It is probable that the Stone Curlew pairs for life, inasmuch that the bird returns annually to favourite haunts, and continues to nest in them season after season. It is not gregarious, although several pairs may frequently be observed nesting within a small area. The nest is invariably on the ground, and consists of a mere hollow scraped out on some bare spot amongst the heather or other vegetation, often on ground strewn with pebbles. No lining appears ever to be inserted in this country, although in India a little dry grass is sometimes used. The bird sits very lightly, running or flying off its eggs at the least alarm, and leaving them to the safety their protective tints ensure.

RANGE OF EGG COLOURATION AND MEASUREMENT: The eggs of the Stone Curlew are invariably two in this country, but Hume states that in India three are sometimes found. They range from clay-colour to yellowish-white in ground, blotched, spotted, or streaked with brown of various shades, sometimes nearly black, and with underlying markings of violet-gray. Two very distinct types are presented, one blotched and spotted with light and dark brown, the other streaked with similar colour; both being marked with gray. On some varieties most of the spots form a zone round the larger end of the egg; on others the markings are evenly distributed over the entire surface. On some the blotches are more or less connected with streaks. A rare variety has few markings of any kind. On some the surface-spots predominate; on others the gray underlying ones are most numerous. Average measurement, 2·1 inches in length, by 1·5 inch in breadth. Incubation, performed by both sexes, lasts, according to Naumann, about seventeen days; other authorities state a month.

DIAGNOSTIC CHARACTERS: The eggs of the Stone Curlew are very characteristic, and can only possibly be confused with those of the Oystercatcher, from which, however, they may be distinguished by their smaller size and paler markings, brown rather than black.

PLATE VI

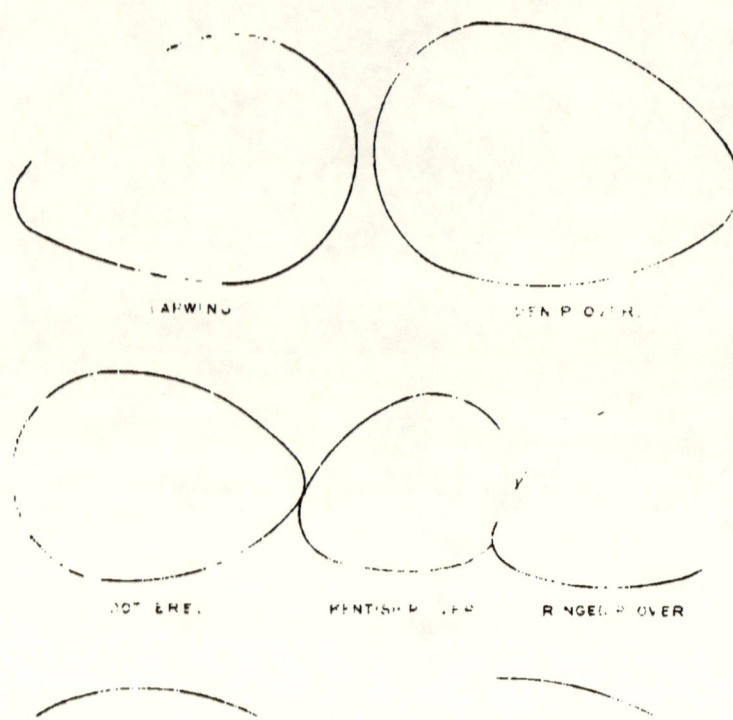

Family CHARADRIID.E. Genus VANELLUS.
Sub-family CHARADRIIN.E.

LAPWING.

VANELLUS CRISTATUS, *Wolf* and *Meyer*.

Single Brooded. Laying season, April and May.

BRITISH BREEDING AREA: The Lapwing is generally distributed throughout the British Islands during the nesting season, breeding in almost every part, including the Hebrides, the Orkneys, and the Shetlands, but certainly preferring high ground for the purpose, especially in the south.

BREEDING HABITS: The Lapwing is a resident in our islands, but its numbers are increased during autumn and winter by arrivals from the Continent, whilst the indigenous birds wander about a good deal, many leaving the more northern and exposed districts entirely, during the latter season. The breeding-haunts of the Lapwing embrace a great variety of scenery. It may be found nesting on moorlands (though not at such an altitude as the Golden Plover), commons, heaths, rough fallows, fields of grain, pastures, and marshes. Although great numbers of birds often breed on a small area of ground, the Lapwing cannot be said to nest in colonies, but it is social enough right through that period. The nest is invariably placed on the ground, either in a little natural hollow, or in the footprint of a cow or horse, or even scraped out by the bird itself. It is frequently under the shelter of a tuft of rushes, or even in the centre; often on the top of a mole-hill, or on the bare turf or ground amongst the growing grain. The hollow is lined with a few bits of dry herbage, but in many cases even this slight provision is omitted. The Lapwing is a light sitter, rising from its eggs at once and flying about the air overhead in a restless, erratic manner, uttering its mewing cry, and generally becoming most demonstrative when furthest from its nest. The eggs are most difficult to find, so closely do they resemble surrounding objects.

RANGE OF EGG COLOURATION AND MEASUREMENT: The eggs of the Lapwing normally are four in number, but instances of *five* have been recorded. They are pear-shaped, and range from buffish-brown of various shades to pale olive and olive-green, rarely to delicate bluish-green in ground colour, richly blotched and spotted with blackish-brown and paler brown, and with underlying markings of ink-gray. Usually most of the markings are large and confluent, and most abundant on the larger end of the egg, the smaller spots being more sparingly distributed over the remainder of the shell. A common variety has the markings principally displayed in a zone round the larger end of the egg; another has them evenly distributed over the entire surface, small and often streaky. Rarely they are very sparingly marked, especially when the ground colour is pale blue. Average measurement, 1·9 inch in length, by 1·3 inch in breadth. Incubation, performed by both sexes, lasts from twenty-five to twenty-six days. If the first clutches of eggs be taken, others will be laid.

DIAGNOSTIC CHARACTERS: The eggs of the Lapwing are very characteristic, and can only be confused in our islands with those of the Golden Plover, from which however they are readily distinguished by their smaller size and browner (not yellow) general appearance.

Family CHARADRIIDÆ.   Genus CHARADRIUS.
Sub-family CHARADRIINÆ.

GOLDEN PLOVER.

CHARADRIUS PLUVIALIS, *Linnæus*.

Single Brooded.   Laying season, May.

BRITISH BREEDING AREA: Next to the Lapwing the Golden Plover is the most widely distributed species of the present sub-family breeding within our limits. It breeds sparingly and locally in Devonshire and Somerset, in various parts of Wales, including the heights of Breconshire, and thence to the moorlands of Derbyshire. From this latter locality northwards along the Pennine area, and throughout Scotland to the Orkneys and Shetlands, and westwards to the Hebrides, it becomes more abundant, breeding commonly in all districts suited to its requirements. In Ireland it is equally common and widely dispersed, breeding on the moors and mountain heaths.

BREEDING HABITS: The Golden Plover is a resident in our islands, but its numbers are increased during winter, and our local birds wander from their upland haunts to the littoral districts at that season. This handsome Plover retires to its breeding-grounds in March and April. These are situated on the upland moors and mountain plateaux, thousands of feet above sea-level in some localities, almost on it in others. The bird cannot be said to nest in colonies, but many pairs may be found breeding within a small area, and throughout the summer it is to a great extent a social, even gregarious species. This Plover appears to pair annually, and generally after arrival at the breeding-grounds. The nest is invariably on the ground, sometimes behind a tuft of cotton-grass, or on a clump of herbage, sometimes amongst short heath, and rarely on bare ground. It is merely a hollow, scantily lined with a few bits of withered herbage or dry grass. The bird is a light sitter, rising from its nest as soon as the moor is invaded, and often seeks to decoy an intruder from the vicinity, or by a nonchalant manner (especially in the male) endeavour to put him off the scent.

RANGE OF EGG COLOURATION AND MEASUREMENT: The eggs of the Golden Plover are four in number, pyriform in shape, and very large for the size of the parent. They are buff of various shades in ground colour, boldly and richly spotted and blotched with dark purplish-brown and blackish-brown, and with a few small underlying markings of gray. Most of the blotches are generally on the larger end of the egg. Average measurement, 2·0 inches in length, by 1·4 inch in breadth. Incubation, performed by both sexes, lasts from sixteen to twenty days.

DIAGNOSTIC CHARACTERS: The eggs of the Golden Plover can only be readily confused with those of the Lapwing in our islands, from which however they are distinguished by the absence of olive (they are a richer buff in general appearance) and their larger size.

Family CHARADRIIDÆ. Genus EUDROMIAS.
Sub-family CHARADRIINÆ.

DOTTEREL.

EUDROMIAS MORINELLUS (*Linnæus*).

Single Brooded.   Laying season, June and early July.

BRITISH BREEDING AREA: The Dotterel is one of the rarest and most local birds that breed within the area of the British Islands. It is more than doubtful whether this species now breeds in any part of England, although formerly it used to do so on many of the chalk ranges in the south, and more recently in the Lake district and on the Cheviots. It now breeds sparingly on the hills of Dumfries-shire, more frequently on the Grampians in North Perthshire, and on the borders of Inverness-shire and Ross-shire. It has been known to nest on the Orkneys. In Ireland it is of only accidental occurrence, never having been known to breed.

BREEDING HABITS: The Dotterel is a summer migrant to our islands, reaching them in small parties towards the end of April or early in May. All through the summer the Dotterel continues more or less gregarious and social, and numbers of nests may be found within a small area of suitable ground. The favourite breeding-haunts of this species are wild uplands and plateaux, the rough, hummocky moorlands or tundras, spread with boulders and clothed with moss, cranberries, and other mountain vegetation. Upon their first arrival the birds are in flocks, but these soon separate more distinctly into pairs and retire to the breeding-places. The nest is invariably placed on the ground, amongst the short moss or grass near the mountain-tops, or on the open moor. Nest it can scarcely be called, for it is simply a hollow amongst the vegetation, with no lining of any kind beyond that which is already in the selected spot. The bird as a rule sits very lightly, leaving the eggs at the first alarm, running along the ridges and occasionally taking a short flight, then returning and standing to watch the intruder; again passing to and fro, and not daring to visit the nest until all but the most untiring patience is exhausted. When the eggs are discovered various alluring antics are frequently practised, and at times, especially if the eggs are much incubated, the parent will remain upon them more closely, then start suddenly up, and reel and tumble as if wounded.

RANGE OF EGG COLOURATION AND MEASUREMENT: The eggs of the Dotterel are always three in number, and vary a good deal in form, some being rotund, others very distinctly pear-shaped, others oval. They vary in ground colour from yellowish-olive to pale buff, richly blotched and spotted with dark brown, and with a few underlying markings of gray. The markings are bold and large, and most numerous on the larger end of the egg, although they are pretty generally dispersed over the entire surface. Average measurement, 1·6 inch in length, by 1·1 inch in breadth. Incubation, performed chiefly by the male, lasts from eighteen to twenty-one days.

DIAGNOSTIC CHARACTERS: The eggs of the Dotterel are very characteristic, and can only be confused with those of the Arctic Tern, but as a rule the underlying markings on those of the latter species are bolder and more numerous. The breeding-grounds of the two species are also widely dissimilar.

Family CHARADRIIDÆ.   Genus ÆGIALOPHILUS.
Sub-family *CHARADRIINÆ*.

KENTISH SAND PLOVER.

ÆGIALOPHILUS CANTIANUS (*Latham*).

Single Brooded.   Laying season, May.

BRITISH BREEDING AREA: The Kentish Sand Plover is one of the rarest birds that breed in the British Islands, and one that will probably soon be utterly exterminated as a nesting species, if the greed of collectors is to be allowed to go on unchecked. Its only nesting-places are on certain parts of the coasts of Kent and Sussex.

BREEDING HABITS: The Kentish Sand Plover is a summer migrant to our islands, arriving towards the end of April or early in May. It is a salt-water species, and frequents sandy beaches intermingled with stretches of shingle during the season of reproduction. It is by no means an unsocial bird, and may be seen in small parties all through the summer, several pairs frequently nesting within a small area of favourable coast. It is not improbable that this species pairs for life, as every year the same favourite spots for nesting are tenanted, and the young and old of a family keep much together during autumn. The Kentish Sand Plover makes no nest, merely laying its eggs in a little hollow amongst the sand or shingle, or on a drift of dry sea-weed and other ocean refuse. The bird sits lightly, leaving its eggs at the least alarm to that safety their protective tints ensure; sometimes feigning lameness, especially if the eggs be near maturity.

RANGE OF EGG COLOURATION AND MEASUREMENT: The eggs of the Kentish Sand Plover are usually three, but frequently four in number. They range from light to dark buff in ground colour, blotched, scratched, and spotted with blackish-brown, and with underlying markings of slate gray. Two distinct types are noticeable. The first and most usual type has the markings in the form of specks and streaks, with a few larger blotches between; the second is more uniformly blotched and spotted, the streaks being not so prominent. Average measurement, 1·2 inch in length, by ·9 inch in breadth. Incubation, performed chiefly by the female, lasts from twenty-one to twenty-three days.

DIAGNOSTIC CHARACTERS: The eggs of the Kentish Sand Plover cannot readily be confused with those of any other species breeding in our islands, the scratchy character of the markings distinguishing them at a glance. They might be confused with some varieties of those of the Lesser Tern, but the markings are always very characteristic, and the shape is constantly more pyriform.

Family CHARADRIIDÆ. Genus ÆGIALITIS.
Sub-family CHARADRIINÆ.

GREATER RINGED PLOVER.

ÆGIALITIS HIATICULA MAJOR (*Tristram*).

Single Brooded. Laying season, middle of April to beginning of June.

BRITISH BREEDING AREA: The large race of the Ringed Plover is widely and generally distributed on the flat sandy coasts of the British Islands, from the Orkneys and Shetlands in the north, the Hebrides in the west, to the Channel Islands in the south. It also frequents the banks of rivers and lochs in many inland localities.

BREEDING HABITS: This Ringed Plover is resident in our islands, but subject to much local movement during the non-breeding season. Its favourite breeding-grounds are long reaches of sandy coast, or the sand-banks and shingly shores of rivers and lakes. Although not exactly breeding in colonies, it remains to a certain extent gregarious during the summer, and numbers of pairs may be found nesting within a small area. Early in April the large flocks break up into smaller parties of paired birds, which retire to their usual nesting-places, the bulk of them finding accommodation on the coast. Rarely the nest, however, may be found at some considerable distance from water of any description. The Ringed Plover makes no nest. In some cases a little hollow is scraped in the sand, but very often even this slight provision is dispensed with. The eggs, however, are always laid well above the usual tide-mark, and on the fine sand rather than on the shingle. The bird sits very lightly. Indeed if the sun shines brightly she is on the eggs but little during the daytime. As soon as the breeding-place is invaded by man the ever-watchful birds slip off their eggs, as a rule manifesting little concern for their safety, seemingly conscious that they are rendered safe by their protective colour, which harmonizes so well with surrounding objects that only a close search can discover them.

RANGE OF EGG COLOURATION AND MEASUREMENT: The eggs of the Ringed Plover are four in number, pyriform in shape, and smooth in texture. They are pale buff or stone-colour in ground, somewhat sparingly spotted and speckled with blackish-brown, and with underlying markings, similar in character, of ink-gray. The spots are generally small and evenly distributed over the surface, but most numerous and largest on the big end of the egg. The range of variation is not very large or pronounced. Average measurement, 1·4 inch in length, by 1·0 inch in breadth. Incubation, performed by both sexes, lasts from twenty-one to twenty-three days.

DIAGNOSTIC CHARACTERS: The eggs of the Ringed Plover cannot easily be confused with those of any other species breeding in our islands, the small, nearly black markings (spots) being very characteristic.

Family CHARADRIIDÆ.   Genus HÆMATOPUS.
Sub-family *TOTANINÆ*.

OYSTERCATCHER.

HÆMATOPUS OSTRALEGUS, *Linnæus*.

Single Brooded. Laying season, May and June.

BRITISH BREEDING AREA: The Oystercatcher, south of Yorkshire and Lancashire, is a somewhat local bird during the breeding season, but north of those localities it becomes much more common, and nests on all parts of the Scottish coasts suited to its requirements, including the Orkneys and Shetlands, the Hebrides, and St. Kilda. In Scotland it also breeds in many inland districts in the courses of the rivers and on the banks of various lochs. It is also widely and generally distributed throughout the coasts of Ireland.

BREEDING HABITS: The Oystercatcher is a resident in our islands, but many birds leave the more northern localities during winter, and its numbers are also increased at that season by individuals from the Continent. The favourite breeding-grounds of this species are stretches of rough pebbles, shingly beaches, low islands, and rock-stacks. The flocks begin to disband in early spring, and to disperse to the breeding-places. We can scarcely regard the Oystercatcher as gregarious in summer, but it is certainly sociable, and numbers of nests may be found at no great distance apart; nevertheless each pair of birds keep a good deal to themselves and to a chosen haunt until the young can fly. I have taken several nests within a few hundred yards, and seen as many as a dozen birds in the air together screaming above their breeding-grounds. The nest of this species scarcely deserves the name. It is little more than a hollow in the shingle, in which the bits of broken shells and pebbles are somewhat neatly arranged. Frequently the eggs are laid on a drift of sea-weed or other ocean refuse. Curious sites are sometimes selected. I have taken the eggs from lofty rock-stacks, and amongst boulders in a little cove, whilst they have been discovered in the deserted nest of a Herring Gull. Usually several mock nests may be found quite close to the one containing the eggs, as if the birds had made several before they were satisfied. The bird sits very lightly, generally rising from the eggs as soon as an intruder is detected, and flying wildly about, uttering their shrill, clear notes. The eggs resemble the surroundings so closely that they are usually found with difficulty, and only after careful search amongst the rougher shingle.

RANGE OF EGG COLOURATION AND MEASUREMENT: The eggs of the Oystercatcher are usually three, sometimes four, and less frequently only two in number. They are pale- or brownish-buff in ground colour, blotched, spotted, and streaked with blackish-brown, and with underlying markings of gray. On some varieties the markings are very streaky; on others they take the form of well-defined spots and small, irregular blotches, either uniformly distributed over the entire surface, or most of them forming a zone round the larger end of the egg. Average measurement, 2·2 inches in length, by 1·5 inch in breadth. Incubation, performed by the female, lasts from twenty-three to twenty-four days.

DIAGNOSTIC CHARACTERS: The eggs of the Oyster-catcher are very characteristic, and as a rule cannot easily be confused with those of any other species breeding in our islands. Some varieties rather closely approach certain types of the eggs of the Stone Curlew, but they are larger, and the spots are always darker and more clearly defined. The breeding-grounds of the two species are also very different.

Family CHARADRIIDÆ. Genus TOTANUS.
Sub-family TOTANINÆ.

RUFF.

TOTANUS PUGNAX (*Linnæus*).

Single Brooded. Laying season, end of May and early June.

BRITISH BREEDING AREA: The Ruff formerly bred commonly in many parts of England, but since the reclamation of so much marsh land it has become very restricted in its distribution. Doubtless incessant persecution by gunners and collectors has had considerable influence in exterminating the Ruff from our shores, and the day is probably not far distant when it will cease to breed within their limits. Ten years ago a female was shot from her nest in Lincolnshire! A few pairs still continue to breed, or attempt to do so, in Norfolk.

BREEDING HABITS: The Ruff is a summer migrant to the British Islands, reaching them towards the end of

April or during the first half of May. Its haunts during the breeding season are in swamps and marshes, wet ground covered with rough hummocks of coarse grass and tufts of sedge rushes and the like. The Ruff is polygamous, one male pairing with several females, and taking no share in nesting duties; consequently we find this species more or less gregarious until the hilling or pairing season is over, when the hens or Reeves go off to incubate their eggs alone. During the mating season the birds congregate at chosen mounds and the males fight for the possession of the females, but as this portion of their economy does not relate very closely to the nest and eggs we may dismiss it without further notice. The nest of the Reeve is made on the ground in the swamps, usually in the centre of a tuft of sedge or coarse grass. It is merely a hollow, lined with a few bits of withered herbage and dead leaves. The female is a close sitter, but is not very demonstrative at the nest.

RANGE OF EGG COLOURATION AND MEASUREMENT: The eggs of the Ruff, or as we might with more propriety say Reeve, are four in number. They vary from greenish-gray to grayish-green in ground colour, spotted and blotched with reddish-brown, and with underlying markings of grayish-brown. As a rule most of the markings are on the larger end of the egg, and they are bolder there than elsewhere and often confluent. Average measurement, 1·7 inch in length, by 1·2 inch in breadth. Incubation, performed by the female, is said by Tiedemann to last sixteen days.

DIAGNOSTIC CHARACTERS: The eggs of the Ruff are characteristic, and not readily confused with those of any other species breeding in the British Islands, with the one possible exception of those of the Redshank. From them, however, they may be distinguished by their grayer or greener ground colour (not so yellow).

Family CHARADRIIDÆ.   Genus TOTANUS.
Sub-family TOTANINÆ.

COMMON SANDPIPER.

TOTANUS HYPOLEUCUS (*Linnæus*).

Single Brooded. Laying season, May and early June.

BRITISH BREEDING AREA: The Common Sandpiper's distribution in our islands is very similar to that of the Ring Ouzel. Commencing in the extreme south-west of England, we find the bird breeding sparingly in Cornwall, Devon, and Somerset, northwards through Wales, where it becomes more abundant, to the Peak district. From this latter locality, northwards, it is common and widely distributed throughout the north of England, and all over Scotland up to the Orkneys and the Shetlands, and westwards to the Hebrides. It is also widely and generally distributed in Ireland.

BREEDING HABITS: The Common Sandpiper is a summer migrant to our islands, reaching them in April or early May; a few individuals, however, winter on our shores, but these may not be birds breeding in our area. The haunts of this lively and engaging little Sandpiper are the gravelly banks of lakes, reservoirs, rivers, lochs, and streams. It is not gregarious nor even social during the breeding season, although several nests may be found within a few hundred yards, each pair, however, keeping to themselves. I am of opinion that the Common Sandpiper pairs for life, and yearly returns to one particular haunt to breed, in spite of continual disturbance. The nest is invariably placed on the ground, and generally, but not always, near the water. A favourite site is on a rough bank clothed with a thin, scattered growth of grass, heath, and other plants; another equally favourite position is beneath a little bush of heath or bilberry, or a tall weed on a bare stretch of sandy ground strewn with pebbles. Instances are on record of its being built in gardens and orchards, in turnip-fields and in woods. The nest is merely a hollow scratched out and lined with scraps of dead heath, withered bents, leaves, and sometimes dry pine needles. The Common Sandpiper sits closely, especially if the eggs are much incubated, and when flushed reels and tumbles along the ground with apparently broken wings to lure the intruder away. Sometimes, however, the sitting bird slips quietly off the nest, running for a few yards before taking wing, as soon as danger threatens, and then the eggs are discovered with difficulty, as they resemble surrounding objects in a very remarkable manner.

RANGE OF EGG COLOURATION AND MEASUREMENT: The eggs of the Common Sandpiper are four in number, and pyriform in shape. They vary from yellowish-white to pale creamy-buff in ground colour, richly blotched and spotted with pale and dark reddish-brown, and with underlying markings of violet-gray. The markings as a rule are not very large, but are usually most abundant on the larger end of the egg. Average measurement, 1·5 inch in length, by 1·1 inch in breadth. Incubation, performed chiefly by the female, lasts about three weeks.

DIAGNOSTIC CHARACTERS: The eggs of this species cannot readily be confused with those of any other bird breeding in our islands, their size, buff ground colour, and comparatively small markings being very characteristic. It is, however, impossible to distinguish certain varieties of the eggs of the Wood Sandpiper from those of the present species, but this bird does not now breed in our islands, as will be seen below.

The Wood Sandpiper (*Totanus glareola*) having once been known with absolute certainty to breed in our islands, calls for some passing notice. Its eggs were obtained by the late Mr. Hancock, forty years ago, on the now drained Prestwick Car, in Northumberland. Its eggs are also reputed to have been taken in Elgin. It is therefore not improbable that odd pairs of this bird may breed from time to time in the British Archipelago. The breeding season of this Sandpiper begins early in May in southern haunts. The nest is usually made on a bit of dryer ground near swamps, in willow thickets, or amongst heath sedge and coarse grass. It is merely a hollow, scantily lined with a few bits of dead herbage. The eggs are four in number, pyriform in shape, and vary from creamy-white or pale buff to very pale olive-brown in ground colour, boldly blotched and spotted with rich reddish-brown, and with a few underlying markings of pale brown. Average measurement, 1·45 inch in length, by 1·0 inch in breadth. It may be remarked that the eggs of the Wood Sandpiper cannot readily be confused with those of any other species breeding in our islands—provisionally, excepting those of the Green Sandpiper, a species that may yet be detected nesting in them. The Wood Sandpiper only rears one brood in the season, and incubation, performed chiefly by the female, lasts about twenty-one days.

PLATE VII

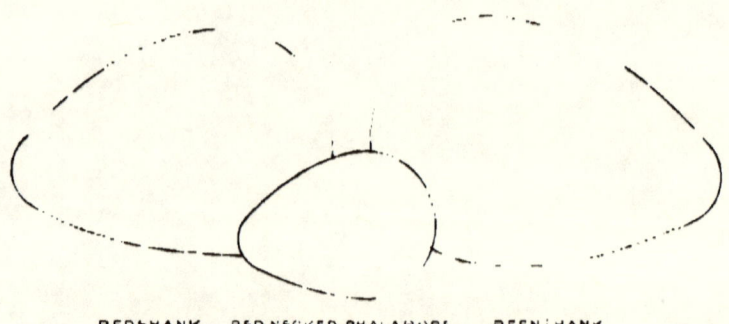

REDSHANK   RED NECKED PHALAROPE   GREENSHANK

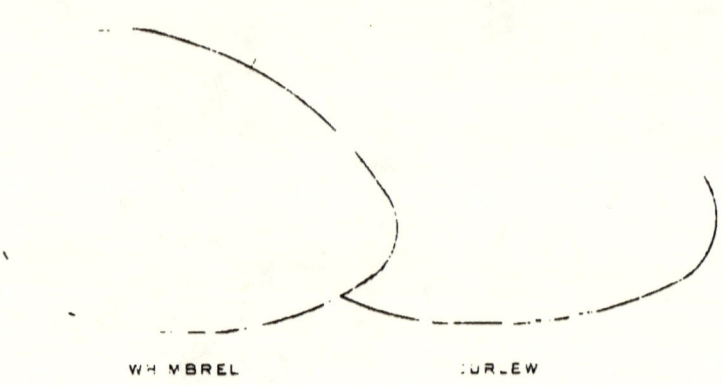

WHIMBREL   CURLEW

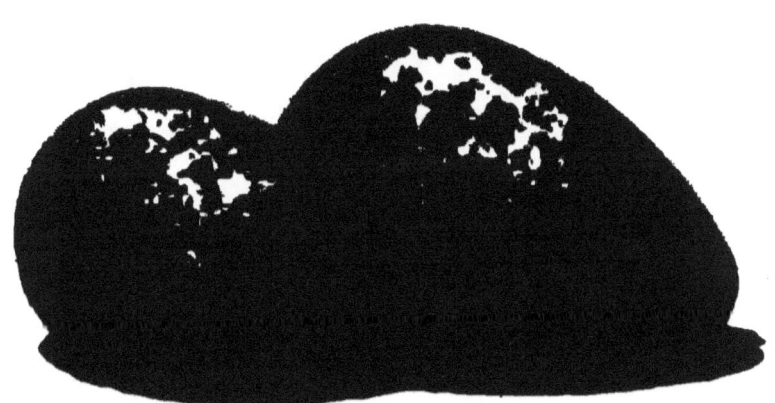

Family CHARADRIIDÆ. Genus TOTANUS.
Sub-family TOTANINÆ.

REDSHANK.

TOTANUS CALIDRIS (*Linnæus*).

Single Brooded.   Laying season, April and May.

BRITISH BREEDING AREA : The Redshank is generally though locally distributed over most of the marshy districts of the British Islands, becoming most common in the eastern counties of England, and even abundant in many parts of Scotland, extending to the Orkneys and the Shetlands, and sparingly to the Outer Hebrides. It is also fairly well distributed over the marshy tracts of Ireland during summer.

BREEDING HABITS : The Redshank is a resident in the British Islands, but subject to considerable local and southern movement during the non-breeding season. Its haunts also vary a good deal according to season, littoral districts being preferred during winter, but the breeding-grounds are more or less inland. Early in spring a movement is made to the nesting-places, which are usually swampy moors, fen and marsh lands, and the boggy shores of mountain lochs. The Redshank is more or less gregarious during the breeding season, and numbers of nests may frequently be found within a small radius of suitable ground. It is probable that this bird pairs for life, as yearly certain haunts will be frequented, and it is also much attached to a favourite site, and has been known to visit it season by season after the district had quite changed in character. During the mating season the male bird is often to be seen trilling high in air, alighting in trees, and displaying its graces in various amatory ways. The nest, if slight, is usually well concealed, and always made upon the ground, often under the shelter of an arched tuft of grass or other herbage, in the centre of a hummock of rushes, or beneath a little bush of heath or a tall weed. The selected site is merely trampled into a slight hollow, and sparingly lined with a few bits of dead vegetation, and often this small provision even is omitted. The bird sits lightly, and when disturbed from the nest often (in company with its mate) becomes very noisy, careering wildly about, or even engages in various alluring actions to decoy an intruder away.

RANGE OF EGG COLOURATION AND MEASUREMENT: The eggs of the Redshank are four in number, and pyriform in shape. They vary from pale buff to dark buff in ground colour, handsomely and boldly blotched and spotted with rich dark brown, and with underlying markings of paler brown and gray. Occasionally a few nearly black streaks occur on the larger end of the egg, where, as a rule, most of the blotches are also displayed. Average measurement, 1·75 inch in length, by 1·2 inch in breadth. Incubation, chiefly performed by the female, lasts about twenty-three days.

DIAGNOSTIC CHARACTERS: The buff ground and large and bold markings distinguish the eggs of the Redshank from those of all other species breeding in our islands with which they are likely to be confused.

Family CHARADRIIDAE. Genus TOTANUS.
Sub-family TOTANINAE.

GREENSHANK.

TOTANUS GLOTTIS (*Linnæus*).

Single Brooded.   Laying season, May.

BRITISH BREEDING AREA: The Greenshank breeds sparingly in the Outer Hebrides, and in a few localities in the inner islands, especially Skye. On the mainland it becomes commoner and more widely dispersed over the counties of Inverness, Argyle, Perth, Ross, Sutherland, and Caithness, but not reaching the Orkneys and the Shetlands. It has never been known to breed in England or in Ireland.

BREEDING HABITS: The Greenshank is a summer migrant to our islands, reaching them towards the end of April, or early in May, but a few are said to spend the winter in Ireland, perhaps individuals that have bred in the Hebrides. Its favourite breeding-grounds are moors, often within sight of the sea, which contain lochs and streams, and abound in bogs and swamps. The Greenshank is neither gregarious nor social during the nesting season, the pairs being scattered here and there over the moors, and each keeping to themselves. It is not improbable that this bird pairs for life, as a considerable attachment to favourite haunts may be remarked. The nest is always made upon the ground, and is a difficult one to find. It is often placed amongst heather or other herbage close to the margin of a stream or loch; often in a tuft of moor grass or on a little mound of dry ground surrounded by swamp. The nest is merely a hollow lined with a few bits of withered herbage. The Greenshank is not a close sitter, rising from the eggs the moment its haunt is invaded, and either flying wildly and noisily about, often in company with its mate, or running restlessly about the moor from hummock to hummock. It will also indulge in various antics to entice an intruder away.

RANGE OF EGG COLOURATION AND MEASUREMENT: The eggs of the Greenshank are four in number, and pyriform in shape. They vary from buffish-white to buff in ground colour, very handsomely blotched and spotted with rich dark brown, and with underlying markings, similar in character, of pinkish-brown and gray. As usual the markings are most numerous and extensive on the larger end of the egg. A somewhat scarce variety is not blotched, but marked with large and small spots over most of the surface; more frequently the markings form an irregular zone round the larger end of the egg. Average measurement, 1·9 inch in length, by 1·35 inch in breadth. Incubation, performed chiefly by the female, lasts about three weeks.

DIAGNOSTIC CHARACTERS: The size and bold dark markings readily distinguish the eggs of the Greenshank from those of allied species breeding in our islands.

Family CHARADRIIDÆ.　　　　　Genus NUMENIUS.
Sub-family *TOTANINÆ*.

COMMON CURLEW.

NUMENIUS ARQUATUS (*Linnæus*).

Single Brooded. Laying season, latter end of April and in May.

BRITISH BREEDING AREA: The Curlew is pretty generally distributed throughout the British Islands in all suitable districts. Its breeding area extends from Cornwall and Devonshire to Somerset, Dorset, Wiltshire, Hants, and most of the uplands of Wales. Thence it extends northwards through the Peak district, Lincolnshire, the entire Pennine Chain (including the Isle of Man), and the Cheviots. Across the Border the bird becomes more abundant, and is widely distributed throughout Scotland, north to the Orkneys and Shetlands, and west to the Outer Hebrides. In Ireland it is equally widely dispersed and common.

BREEDING HABITS: The Curlew is a resident in the British Islands, but like many other species it changes its ground a good deal with the season, is subject to much local movement, and its numbers are increased in winter by arrivals from the Continent. The breeding-grounds of the Curlew are moorlands, especially those of a swampy nature, and at a considerable distance above sea-level, rough unenclosed mountain pastures, and arable uplands. In March or early in April the Curlews begin to leave the coasts, and to return to their breeding-haunts, pairing, and scattering themselves up and down the moors and rough lands. The nest is invariably made on the ground, generally on some dry part of the moor, under the shelter of a bush, or in the centre of a tuft of grass or rushes, but occasionally the eggs are laid on the rough fallows without nest of any kind. The nest is a mere shallow hollow, sparingly lined with a few bits of withered herbage or dead leaves. Although the Curlew can scarcely be regarded as gregarious at this season, numbers of birds often breed within comparatively small areas, and when one pair is disturbed the entire locality is soon in commotion. The bird sits lightly, rising from the nest at the first alarm, often given by its watchful mate, and becomes noisy enough as it flies about in alarm.

RANGE OF EGG COLOURATION AND MEASUREMENT: The eggs of the Curlew are four in number, usually pyriform, but sometimes rotund in shape and large for the size of the bird. They vary from olive-green to buff in ground colour, blotched and spotted with olive-brown, and with underlying markings of pale gray. Occasionally a few streaks of blackish-brown occur. Several types are presented. One variety has the markings comparatively small, and mostly congregated in a zone round the larger end of the egg; another has the markings uniformly distributed over the entire surface; another has the markings large and well-defined, few, and very rich in colour. Average measurement, 2·7 inches in length, by 1·85 inch in breadth. Incubation, performed chiefly by the female, lasts about thirty days.

DIAGNOSTIC CHARACTERS: The large size, pyriform shape, and olive appearance distinguish the eggs of the Curlew from those of allied species breeding in our islands.

Family CHARADRIIDÆ. Genus NUMENIUS.
Sub-family TOTANINÆ.

WHIMBREL.

NUMENIUS PHÆOPUS (*Linnæus*).

Single Brooded. Laying season, May and June.

BRITISH BREEDING AREA: The Whimbrel's breeding area in the British Islands is a remarkably restricted one, and is nowhere known to extend on to the mainland. A few pairs nest on North Ronay in the Hebrides, and on the Orkneys, whilst a greater number do so on the Shetlands. It is not improbable that this species breeds elsewhere in the Hebrides; I saw it on St. Kilda during June.

BREEDING HABITS: The Whimbrel is a summer migrant to the British Islands, reaching our shores in small numbers in April, but passing over them much more abundantly in May. Its favourite breeding-grounds are the wild, elevated moorlands at no great distance from the sea. The bird probably pairs each season, although I may remark that the information on this point is very unsatisfactory. It is not a gregarious species, but numbers of pairs often nest in a small area, just like the Curlew. The nest is always placed on the ground amongst heath, or beneath the shelter of a tuft of grass in a dry part of the swampy moor. It is merely a hollow, scantily lined with a few bits of withered herbage or dead leaves. The actions of the Whimbrel at the breeding-grounds are very similar to those of the Curlew, and the notes of both species are much the same.

RANGE OF EGG COLOURATION AND MEASUREMENT: The eggs of the Whimbrel are four in number, and pyriform in shape. They very closely resemble those of the Curlew in general appearance, and run through pretty much the same range of variation. They vary from olive-green to buff in ground colour, blotched and spotted with olive-brown and reddish-brown, and with underlying markings of pale gray. On some eggs the spots are mostly distributed in a zone round the larger end; on others they are evenly dispersed. Average measurement, 2·3 inches in length, by 1·6 inch in breadth. The period of incubation appears to be unknown, but is probably the same as that of the Curlew, and is performed chiefly by the female.

DIAGNOSTIC CHARACTERS: The small size readily distinguishes the eggs of the Whimbrel from those of the

Curlew, which are always more bulky. They somewhat closely resemble certain varieties of those of Richardson's Skua, but may be distinguished from them by their slightly larger size and much more pyriform shape.

Family CHARADRIIDÆ.　　　　　　Genus PHALAROPUS.
Sub-family TOTANINÆ.

RED-NECKED PHALAROPE.

PHALAROPUS HYPERBOREUS (*Linnæus*).

Single Brooded.　Laying season, end of May and first half of June.

BRITISH BREEDING AREA: "Once upon a time" the Red-necked Phalarope bred in the counties of Sutherland, Inverness, and Perth, but the bird now nests nowhere on the mainland of the British Isles. Its breeding area is confined to the Shetlands, the Orkneys, and some few of the Outer Hebrides—North and South Uist, Benbecula, etc. Even here it is a rare and remarkably local bird, and destined to ultimate extermination if steps are not taken to protect it from the persecutions of oologists and their jackals the trader collectors.

BREEDING HABITS: The Red-necked Phalarope is a summer migrant to its breeding-grounds in our islands, reaching them towards the end of April or early in May. Its breeding-places are on moors studded with rush-fringed pools at no great distance from the sea. It is probable that this species pairs for life, as it yearly returns to the same places to breed. It is also a gregarious bird during the breeding season, and its nests are placed in more or less scattered colonies. The nest is slight, and usually placed on the ground, in our islands, amongst the grass or other herbage on a patch of dry ground in the marshes and close to the pools. Sometimes it is placed in the centre of a grass or rush-tuft. It is merely a hollow scantily lined with a few bits of dry grass and broken rush. The Red-necked Phalarope is remarkably tame and confiding at its nest, leaving it when disturbed, and usually flying to the nearest water, evincing little or no anxiety for its safety.

RANGE OF EGG COLOURATION AND MEASUREMENT: The eggs of the Red-necked Phalarope are four in number, and pyriform in shape. They vary in ground colour from pale olive to buff of various shades, blotched and spotted with umber-brown, blackish-brown, and pale brown, and with a few underlying markings of gray. As is usual, most of the blotches are on the larger end of the egg. Average measurement, 1·1 inch in length, by ·82 inch in breadth. Incubation, performed chiefly by the male, lasts about three weeks.

DIAGNOSTIC CHARACTERS: The small size, pyriform shape, and colour combined, readily distinguish the eggs of the Red-necked Phalarope from those of any other allied species breeding in the British Islands.

Family CHARADRIIDÆ. Genus TRINGA.
Sub-family SCOLOPACINÆ.

DUNLIN.

TRINGA ALPINA, *Linnæus*.

Single Brooded. Laying season, early May and June.

BRITISH BREEDING AREA: The Dunlin is the only species in the present genus that breeds within our area. It nests locally and sparingly in Cornwall, Devon, and Somerset (possibly on the Welsh mountains), the marshes of the Dee, Lancashire, Yorkshire, and more commonly in Cumberland, Northumberland, and throughout the west of Scotland—including the Hebrides—north to Sutherlandshire, the Orkneys and Shetlands. In Ireland it breeds somewhat sparingly and locally in the north-west.

BREEDING HABITS: The Dunlin is a resident in our islands subject to considerable local movement, and its numbers are largely increased in autumn by birds from other lands. Its breeding-grounds in this country are marshy woodlands and mountain swamps at no great distance from the sea as a rule, or at least within a comparatively short distance of tidal waters. During March and April the birds leave their winter haunts on the coasts and mud-flats, and retire to the breeding-grounds. Although gregarious enough at all other times of the year, in the nesting season the bird can scarcely be considered so. It is, however, social even then, and numbers of pairs may be found breeding within a small area of suitable ground. The bird appears to pair annually, and during that period the males soar and trill. The nest, invariably on the ground, well concealed and difficult to find, is usually placed in a tussock of grass or rushes, or beneath a little bush of heather or bilberry, or even amongst patches of thrift on bare sandy soil. It is a slight structure, a mere hollow lined with a few dead leaves and bits of dry grass, with perhaps a few twigs or roots round the margin. The bird is a close sitter, but when disturbed will often engage in alluring antics to arrest attention from its eggs.

RANGE OF EGG COLOURATION AND MEASUREMENT: The eggs of the Dunlin are four in number, and pyriform in shape. They vary from pale olive to pale brown and buff in ground colour, handsomely blotched and spotted with rich reddish and blackish brown, and with a few obscure underlying markings of gray. They are generally very boldly marked, especially on the larger end, where the blotches and splashes frequently become confluent and hide most of the ground colour. Occasionally a few nearly black streaks occur on the larger end of the egg: a less frequent variety has the spots smaller and more evenly distributed over the entire surface. Average measurement, 1·3 inch in length, by ·95 inch in breadth. Incubation, performed by the female, lasts twenty-one or twenty-two days.

DIAGNOSTIC CHARACTERS: The eggs of the Dunlin cannot be easily confused with those of any other allied species breeding in the British Islands, being readily distinguished by their size and their handsome, well-marked appearance. Some eggs closely resemble those of the Common Snipe, but their size prevents any possibility of confusion.

Family CHARADRIIDÆ.   Genus SCOLOPAX.
Sub-family SCOLOPACINÆ.

WOODCOCK.

SCOLOPAX RUSTICOLA, *Linnæus*.

Probably Single Brooded. Laying season, March and especially April; May.

BRITISH BREEDING AREA: The Woodcock breeds sparingly throughout the British Islands, wherever suitable cover is to be found. It probably nests much more abundantly than is generally supposed, owing to its retiring habits and the nearly entire absence of observation during its breeding season.

BREEDING HABITS: The Woodcocks that breed in our islands are undoubtedly resident therein, but numbers of birds visit our shores in autumn from other lands, and numbers pass over them during the two seasons of passage. The breeding-haunts of this species are plantations of young trees, spinneys, and woods in which plenty of bottom growth and long rank vegetation clothes the ground. The Woodcock pairs annually, and during the period of its "roding" or mating flights in spring, is to a certain extent social; otherwise this species is solitary, each pair for the rest of the nesting time keeping to themselves, although several nests may be found at no great distance apart. The nest is always made on the ground, in a dry secluded corner of the wood or plantation, where plenty of cover may be found in thickets of bracken, fern, brambles, dry grass, and drifts of autumn leaves. Sometimes a bare situation at the foot of a tree is selected. The nest is a mere hollow, lined with dry grass and dead leaves, and is usually well concealed by surrounding vegetation. The lining is sometimes increased whilst incubation is in progress, especially when the nest is in a rather exposed site, as if the bird was anxious to assimilate itself with the tints of surrounding objects as closely as possible. To further these designs the Woodcock is a close sitter, remaining brooding over the nest until flushed, as if conscious that it was unseen and could not easily be detected in a spot where its rich brown dress harmonizes so closely with the ground around it. It is probable that the eggs are covered when the sitting bird leaves the nest voluntarily.

RANGE OF EGG COLOURATION AND MEASUREMENT: The eggs of the Woodcock are four in number, and nothing near so pyriform in shape as is usually the case amongst this family of birds. They vary in ground colour from very pale yellowish-brown to buffish-brown, rather sparingly spotted and blotched with reddish-brown, and with similar underlying markings of gray. They are not subject to any very great range of variation in colour. Average measurement, 1·7 inch in length, by 1·35 inch in breadth. Incubation, performed chiefly by the female, lasts about three weeks.

DIAGNOSTIC CHARACTERS: The size, form, pale colouration, and smallness of the spots (not blotches), readily distinguish the eggs of the Woodcock from those of any other allied species breeding in our area.

Family CHARADRIIDÆ.   Genus SCOLOPAX.
Sub-family SCOLOPACINÆ.

COMMON SNIPE.

SCOLOPAX GALLINAGO, *Linnæus*.

Single Brooded. Laying season, latter half of April to middle of May.

BRITISH BREEDING AREA: The Common Snipe is very generally distributed over the British Islands, breeding in all suitable localities, but more abundantly in Scotland and Ireland than in England.

BREEDING HABITS: The Common Snipe is a resident in the British Islands, but subject to much local movement during the non-breeding season, and largely increased in numbers in autumn and winter by birds from other lands. The favourite breeding-grounds of this species are marshes, wet moorlands, and boggy ground in open country. It is not a gregarious bird, each pair keeping to one particular haunt, although many nests may be found within small areas of suitable ground. Space will not allow of any detailed description of the drumming or bleating of the male bird during the pairing season; it is a sound made whilst the bird is soaring above its breeding-haunts, indulging in those aërial flights peculiar to this group. The Common Snipe pairs annually. The nest is generally made in the centre, or by the sheltering side of a tuft of rushes or coarse grass in the swamps, and is a slight hollow lined with dry grass and bits of dead aquatic herbage. The bird sits closely, usually remaining on the nest until nearly trodden upon, and makes little or no demonstration when flushed.

PLATE VIII.

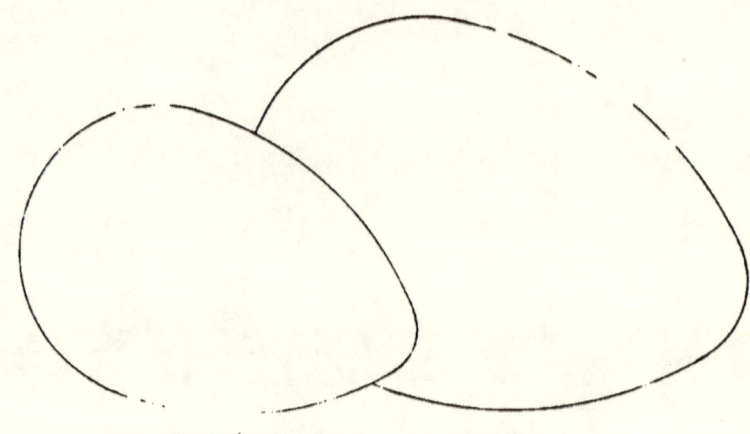

RICHARDSON'S SKUA.   GREAT SKUA.

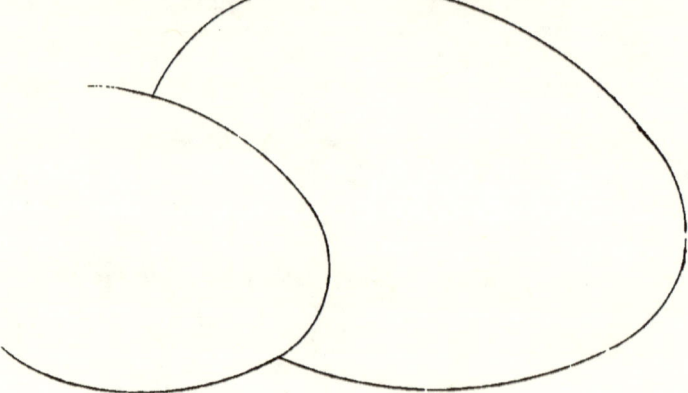

KITTIWAKE.   GREAT BLACK BACKED GULL.

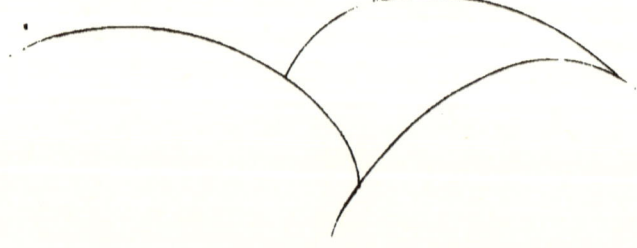

RANGE OF EGG COLOURATION AND MEASUREMENT: The eggs of the Common Snipe are four in number, and pyriform in shape. They vary from buff of various shades to olive of various shades in ground colour, handsomely and boldly blotched and spotted with rich dark brown, occasionally streaked with blackish-brown, and with numerous large and small underlying markings of pale brown and gray. As is usual, the larger end of the egg is most richly marked, and the spots not unfrequently form a zone round it. Average measurement, 1·6 inch in length, by 1·1 inch in breadth. Incubation, performed by the female, lasts from sixteen to twenty days.

DIAGNOSTIC CHARACTERS: The size, shape, and handsome appearance of the eggs of the Snipe readily distinguish them from those of any allied species breeding in our islands.

Family STERCORARIIDÆ.   Genus STERCORARIUS.

RICHARDSON'S SKUA.

STERCORARIUS RICHARDSONI (*Swainson*).

Single Brooded.   Laying season, latter end of May and early June.

BRITISH BREEDING AREA: Richardson's Skua is another of our local species, breeding sparingly in Caithness and Sutherlandshire, on the Hebrides, and more commonly on the Orkneys and Shetlands.

BREEDING HABITS: Richardson's Skua is a summer migrant to our islands, but best known as passing them on passage during April and May. It arrives at its British breeding-places early in the latter month. The breeding-haunts of this Skua are wild open moorlands at no great distance from the sea—vast expanses of heath, and rank grass and moss, broken up into marshes and studded with pools. It is more or less social during the summer, but does not breed exactly in colonies, although many pairs may be scattered over the moors in one locality. It appears to pair annually. The nest is invariably made on the ground, with little or no attempt at concealment. It is merely a hollow, lined with a little dry grass or other withered herbage; and in some cases is nothing but a depression in the moss. The bird sits lightly, leaving the eggs as soon as the breeding-ground is invaded by man, and flying to and fro, sometimes swooping within a few inches of his head, and frequently betraying their whereabouts by too much anxiety for their safety. If care be taken not to alarm the bird more, it usually soon settles on the ground again, and eventually returns to the nest.

RANGE OF EGG COLOURATION AND MEASUREMENT: The eggs of Richardson's Skua normally are two in number, but instances of three having been found in one nest are on record, whilst occasionally only one is found. They vary considerably in shape, some being very pyriform and pointed, others more rotund. They vary in ground colour from olive to brown of various shades, spotted and speckled with very dark brown, and with a few obscure underlying markings of grayish-brown. Some varieties have many of the spots more or less elongated into comma-shaped marks, or irregular masses; others have most of the markings congregated in a zone round the larger end of the egg; more generally they are pretty evenly distributed over the entire surface. Average measurement, 2·3 inches in length, by 1·6 inch in breadth. Incubation, performed by the female, lasts about a month.

DIAGNOSTIC CHARACTERS: The eggs of this Skua require the most careful identification, as they may easily be confused with those of the Common Gull, which breeds in similar localities. They also resemble very closely certain varieties of those of the Black-headed Gull; the locality is of the first importance in deciding their authenticity, if not taken from the nest in the presence of the parents.

Family STERCORARIIDÆ. Genus STERCORARIUS.

GREAT SKUA.

STERCORARIUS CATARRHACTES (*Linnæus*).

Single Brooded. Laying season, May.

BRITISH BREEDING AREA: The Great Skua is another excessively local species in the British Islands, its only nesting-places being in the Shetlands, small colonies breeding on Unst and Foula, the former the most northerly of the group, and the latter some twenty miles to the westward of Mainland. It is a source of satisfaction to know that the birds are jealously preserved from extermination; were it otherwise a few years would suffice to banish this fine bird from the list of our breeding species.

BREEDING HABITS: The Great Skua is best known on our shores as a coasting migrant on its way to or from more northern breeding-haunts. By the end of April the few pairs that nest within our limits betake themselves to their breeding-places on the high moorlands. It is not improbable that this species pairs for life, although the sexes do not keep very close company until spring. It is social during the breeding season, nesting in scattered colonies, the pairs being distributed over a considerable area of moor. The nest is invariably on the ground, and in most cases is little more than a hollow in the moss, sometimes lined with a few bits of dry grass. The Great Skua is not a close sitter, rising from its eggs as soon as its haunts are invaded, and swooping boldly round the head of the intruder, courageously endeavouring to drive him from the sacred spot.

RANGE OF EGG COLOURATION AND MEASUREMENT : The eggs of the Great Skua are two in number. They vary from pale buff to dark buffish-brown or olive-brown in ground colour, somewhat obscurely spotted and speckled with dark brown, and with underlying markings of grayish-brown. As a rule the spots are most numerous, and many of them often confluent, on the larger end of the egg. Average measurement, 2·9 inches in length, by 2·0 inches in breadth. Incubation, performed chiefly by the female, lasts about a month.

DIAGNOSTIC CHARACTERS : As this species is so very local during the nesting season, its eggs cannot readily be confused with those of any other species breeding within our area. Away from the nest, however, they cannot always with absolute certainty be distinguished from those of the Herring Gull and the Lesser Black-backed Gull. A seldom-failing point of distinction, however, is the much more obscure and ill-defined spots on those of the Great Skua.

Family LARIDÆ. Genus LARUS.
Sub-family LARINÆ.

KITTIWAKE.

LARUS TRIDACTYLUS, *Linnæus.*

Single Brooded.   Laying season, May and June.

BRITISH BREEDING AREA: The Kittiwake is widely distributed along the most rocky coasts of the British Islands, becoming, however, more local in England than elsewhere. It breeds at Flamborough Head and the Farne Isles on the east coast; in the south only on the coasts of Devon and Cornwall, the Scilly Isles and Lundy; more numerously on certain parts of the Welsh coast, less so on the Isle of Man. From the nature of the coast, Scotland is more favourable to the requirements of this species, and in many places it breeds in enormous colonies, especially on the west, and on the Hebrides (including St. Kilda), the Orkneys, and the Shetlands. On the east coast of Scotland, however, it is not so widely dispersed, but breeds on the Bass, the May, and in Aberdeenshire. In Ireland it is equally widely distributed on all rocky coasts suited to its requirements.

BREEDING HABITS: The Kittiwake is a resident in our islands, but subject to much local movement during the non-breeding season, being then more widely dispersed and more oceanic in its habits. The breeding-places of the Kittiwake are lofty wall-like ocean cliffs and rock-stacks. It is a most gregarious species, and where the accommodation is ample its colonies are of enormous dimensions, many thousands of birds in some places breeding in company. On the other hand, in less suitable places, only a few pairs may be met with. Early in spring the birds begin to arrive at their breeding-places. As these, and even the old nests in many cases, are returned to each season, it is probable that this Gull pairs for life. The nests are usually made on ledges, in crevices, and on projections and buttresses of the most inaccessible cliffs; in large colonies every possible site is utilized, many of them being side by side. They are made at various heights from the sea, sometimes as low as ten feet, but more often midway up the rocks. The nest is large and well made. Externally it is made of turf and roots, with the soil adhering caked and matted together. Upon this a further nest of dry sea-weed and stalks of marine plants is formed, lined with dry grass, and occasionally a few feathers. The nest and the rocks near it are usually well whitewashed with droppings. The bird sits rather closely, as if conscious of its safety in its inaccessible haunt, but when disturbed flies restlessly about, uttering its noisy cry. The din made by a colony of disturbed Kittiwakes must be heard to be realized.

RANGE OF EGG COLOURATION AND MEASUREMENT: The eggs of the Kittiwake are usually two or three, more rarely four in number. They vary from greenish-blue and brownish-olive to pale buff and buffish-brown in ground colour, blotched and spotted with reddish-brown, and with underlying markings of paler brown and gray. One variety has the markings few and large; another is zoned round the larger end with smaller spots and blotches; another is boldly blotched over the entire surface; on another the markings take the form of short, irregular streaks. The pale underlying markings are both large and numerous, and on certain types preponderate over the surface-markings, which are small and indistinct. Average measurement, 2·15 inches in length, by 1·6 inch in breadth. Incubation, performed chiefly by the female, lasts twenty-six days.

PLATE IX

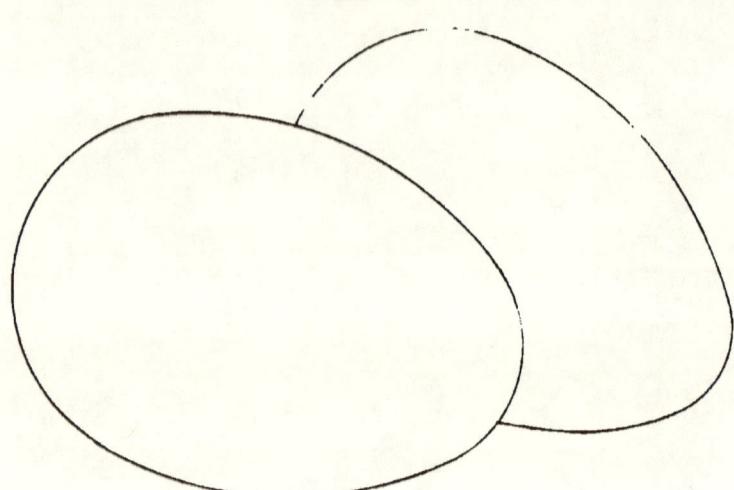

HERRING GULL.   LESSER BLACK BACKED GULL.

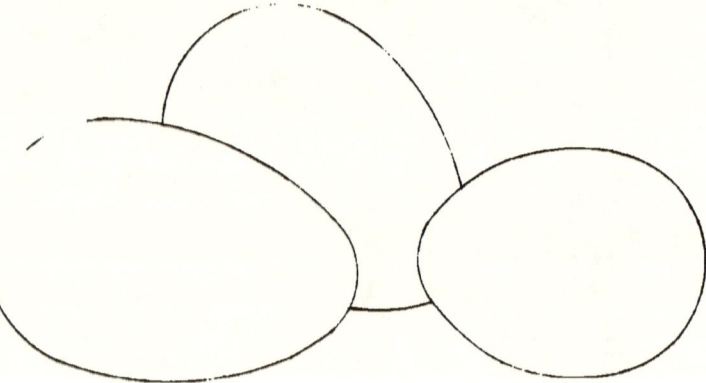

SANDWICH TERN   COMMON TERN.

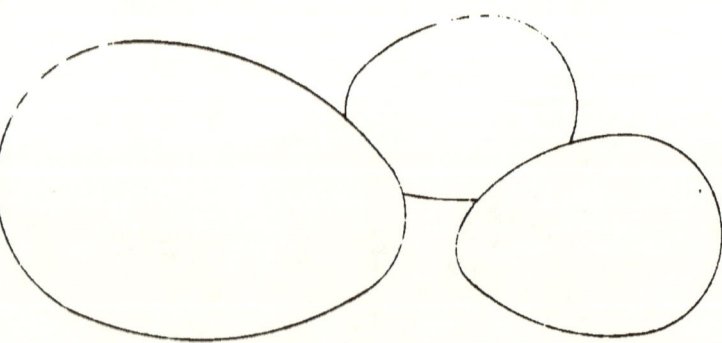

BLACK GUILLEMOT.   LESSER TERN   ARCTIC TERN

DIAGNOSTIC CHARACTERS: The size, general brown appearance, and large character of the markings distinguish the eggs of the Kittiwake from those of other species. When taken from the nest no confusion can possibly arise.

Family LARIDÆ. Genus LARUS.
Sub-family *LARINÆ*.

HERRING GULL.

LARUS ARGENTATUS, *Gmelin*.

Single Brooded. Laying season, May and early June.

BRITISH BREEDING AREA: The Herring Gull is widely distributed round the British coasts, from the cliffs of the south coast of England to those of the Orkneys and the Shetlands, from the Farne Islands and the Bass in the east to the Hebrides, St. Kilda, the Irish mainland, and the Blaskets in the west.

BREEDING HABITS: The Herring Gull is a resident in the British Islands, but more widely distributed in winter than in summer. Its principal breeding-places are on ocean cliffs, both those that are broken up into crags and downs, and those that rise sheer from the water; rock-stacks, and low rocky islands. Much less frequently it resorts to marshes (Foulshaw Moss in Westmoreland, for instance) and small islands in lochs. This bird is gregarious during the breeding season, and in some localities, as for instance at the Farnes, and on the cliffs between Berry Head and Dartmouth, its colonies are very extensive. Owing, however, to the ease with which this Gull accommodates itself to circumstances, its breeding area is wide and varied, not confined to one or two chosen localities, as is the case with so many other species. Early in spring the birds return to their nesting-places, but the breeding season does not commence before May. The nests are built in a great variety of situations. In some cases they are made on ledges and in crevices of the cliffs, or amongst crags or in sheltered hollows of the grassy downs; in others they are built on the summit of rock-stacks, or amongst the dense growth of sea-campion and thrift, as at the Farne Islands. When on the cliffs the nest is usually larger than when on the ground or amongst crags. It is composed of turf, dry sea-weed, straws, and stalks of marine plants, and lined with grass, much of it often semi-green. Sometimes the entire nest is merely composed of a few straws and bits of dry grass. I have half a dozen, taken from near Berry Head, that could all be grasped in one hand. When the breeding-place is invaded the birds become very clamorous, and fly to and fro in alarm, rising from their nests and settling again at the first opportunity. When the nests are on cliffs the birds usually leave them with reluctance.

RANGE OF EGG COLOURATION AND MEASUREMENT: The eggs of the Herring Gull are two or three in number. They vary in ground colour from pale bluish-green to olive-brown and yellowish-brown, somewhat sparingly spotted with dark brown, and with underlying markings of paler brown and gray. The markings are usually in the form of spots rather than blotches, but on some eggs they take the latter character. Average measurement, 2·9 inches in length, by 2·0 inches in breadth. Incubation, performed by both sexes, lasts twenty-six days.

DIAGNOSTIC CHARACTERS: As a rule the eggs of this Gull may be distinguished by their size and the markings being in the form of spots rather than blotches, but they require careful identification, especially in localities where the Lesser Black-backed Gull breeds in company with this species.

Family LARIDÆ.
Sub-family LARINÆ.

Genus LARUS.

GREAT BLACK-BACKED GULL.

LARUS MARINUS, *Linnæus*.

Single Brooded. Laying season, May and early June.

BRITISH BREEDING AREA: The Black-backed Gull breeds nowhere along the eastern coast-line of England, and only sparingly on the south coast in Dorset; on the west it does so in Cornwall and the Scilly Isles, Lundy, here and there on the Welsh coast, and probably in the Solway district. In Scotland, however, it becomes commoner and more generally dispersed, especially along the west coast, notably in the Hebrides, including St. Kilda, and on the north, including the Orkneys and the Shetlands. It is also widely dispersed in Ireland, in districts suited to its habits.

BREEDING HABITS: The Great Black-backed Gull is a resident in our islands, but more widely dispersed in winter than in summer. Its favourite breeding-places are ranges of ocean cliffs, more especially rock-stacks, and less frequently small islands in mountain lochs, even at some distance inland. In some places it may be met with breeding in small colonies, but more usually in scattered pairs. This fine Gull probably pairs for life, and may be found breeding year after year in one particular spot. The nest is generally made on a stack of rocks, or a ledge of the cliffs; less frequently on the ground, on an island. It is slight and loosely put together, a mere hollow in most cases, carelessly lined with grass, twigs, dry sea-weed, or stalks of marine vegetation. I have known the eggs to be laid on the bare ground, in a hollow amongst the crags. The birds sit lightly, but are bold and clamorous when disturbed from the nest.

RANGE OF EGG COLOURATION AND MEASUREMENT: The eggs of the Great Black-backed Gull are usually three in number, less frequently two, and rarely even one. They vary from grayish-brown to brown tinged with olive in ground colour, spotted with dark umber-brown, and with underlying markings of brownish-gray. As a rule the markings are not very numerous, range in size from that of a buckshot downwards to a speck, and are distributed over most of the surface. Some varieties, however, are most heavily marked on the larger end. Average measurement, 3·1 inches in length, by 2·1 inches in breadth. Incubation, performed by both sexes, lasts about a month.

DIAGNOSTIC CHARACTERS: The large size and small markings readily distinguish the eggs of this Gull from those of allied species breeding in our islands.

Family LARIDÆ. Genus LARUS.
Sub-family *LARINÆ*.

LESSER BLACK-BACKED GULL.

LARUS FUSCUS, *Linnæus*.

Single Brooded. Laying season, May and June.

BRITISH BREEDING AREA : The Lesser Black-backed Gull is very commonly and widely distributed throughout the British Islands in all suitable districts. These, however, nowhere occur in the east of England south of the Tyne and east of Devonshire. It breeds in Devon, Cornwall, Wales, the Isle of Man, Cumberland, Northumberland (including the Farne Islands), and in Scotland up to the Orkneys and the Shetlands in the north, and to the Hebrides in the west. In Ireland it is extremely local, and only breeds in one or two localities.

BREEDING HABITS : The Lesser Black-backed Gull is a resident in the British Islands, but subject to much local movement during the non-breeding season, and most of the birds inhabiting the extreme northern districts draw south in winter. Its breeding-grounds embrace a great variety of situations. The favourite haunts are low rocky islands which contain plenty of herbage ; rock-stacks, and islands in inland lakes, grassy downs, mosses, and flows. This Gull is remarkably gregarious, and usually breeds in more or less extensive colonies—one of the most extensive known to me being on the Farne Islands. The bird may probably pair for life, as season by season the same breeding-place is used. The nest of this species varies a good deal in size, even in the same colony of birds. As a rule it is a somewhat bulky structure, often placed amongst a bed of campion or thrift, made of turf, branches of heather, leaves and stalks of marine vegetation, and sea-weed, and lined with grass, much of it semi-green. The poorer nests are usually made in hollows of the downs, or in ledges of crags, or amongst broken rock, and these are mere depressions more or less sparingly lined with grass and perhaps one or two feathers. When the colony is invaded by man the birds rise in crowds from their nests and circle overhead, uttering their incessant cries. Few sights are grander than the colony of these birds at the Farne Islands, the gulls filling the air literally like a heavy snowstorm.

RANGE OF EGG COLOURATION AND MEASUREMENT: The eggs of the Lesser Black-backed Gull are usually three in number, but sometimes four. They vary in colour to an astonishing degree. I made the following note on their colour from a heap of many thousands: pale green, bluish-white, dark olive-brown, pale brown, buff, and gray constitute the ground colours, and the spots and blotches vary from dark liver-brown to pale brown and gray. Some eggs are streaked almost like those of a Bunting, others are finely marked over the entire surface; whilst others yet again are boldly blotched, or have most of the markings in a zone round the larger end. Average measurement, 2·7 inches in length, by 1·9 inch in breadth. Incubation, performed by both sexes, lasts about a month.

DIAGNOSTIC CHARACTERS: Unfortunately I know of no absolutely reliable character by which the eggs of this Gull can be distinguished from those of the Herring Gull, or from certain varieties of those of the Common Gull. Those of the former species, however, are generally larger, those of the latter smaller, whilst on both the markings generally take the form of *spots* rather than blotches. Careful identification, however, is necessary.

Family LARIDÆ. Genus LARUS.
Sub-family LARINÆ.

COMMON GULL.

LARUS CANUS, *Linnæus*.

Single Brooded. Laying season, latter half of May and early June.

BRITISH BREEDING AREA: The Common Gull has little claim to its trivial name, for it is one of our most local birds during the breeding season. It is not now known to breed anywhere in England, although it formerly did so in Lancashire. In Scotland, however, it nests in the Solway district, and thence northwards, both on the coasts and in many inland districts, it is widely distributed, reaching the Orkneys and the Shetlands in the north and the Hebrides in the west, but becoming rarer in the east. In Ireland it is somewhat local, but certainly breeds in Counties Donegal, Sligo, and Mayo, and on the Blaskets, off the coast of Kerry.

BREEDING HABITS: The Common Gull is a resident in the British Islands, but subject to much local and southern movement after the breeding season. Its breeding-places are varied, and occur both inland and on the coast. Sometimes a locality is chosen on a small island in a mountain lake, or the summit of a stack of rocks, or a marshy tract on the banks of a lake, or even grassy downs by the sea. But the most extensive colonies of this Gull that I have ever seen in our islands were situated on rocky islets in deep sea-water lochs in the Hebrides. In some of these wild, secluded lochs almost every island contained a colony, some of only a few pairs, others larger. The main colony was on a rather low island, which sloped up from a sandy beach, and fell here and there in little precipices to the sea. It is probable that this species pairs for life. The Common Gull is gregarious during the breeding season, but the colonies vary considerably in size, and many pairs are scattered about in some localities where the bird is not common, or suitable sites are scarce. Late in April the birds congregate at their breeding-stations, and nest-building soon after commences. The nests are placed in various situations. Some are made amongst crevices of the rocks, others amongst grass and heather; some in hollows in the bare turf, others on ledges of cliffs. In Norway this Gull has been known to lay in the deserted nest of a Hooded Crow in a pine tree. The nest is made of heather branches, turf, dry grass, or sea-weed and stalks of marine plants, and lined with grass, in some cases almost green: occasionally the nest is little more than a mere hollow. When disturbed at their breeding-place the birds become very clamorous, and fly to and fro in alarm.

RANGE OF EGG COLOURATION AND MEASUREMENT: The eggs of the Common Gull are usually three in number, but sometimes four. They vary in ground colour from olive-brown to buffish-brown, spotted and often streaked with dark brown, and with underlying markings of brownish-gray. As usual, most of the markings (from the size of buckshot downwards) are on the larger end of the egg, where they sometimes form an irregular zone, but some varieties are more evenly spotted over the entire surface. Average measurement, 2·25 inches in length, by 1·6 inch in breadth. Incubation, performed by both sexes, lasts about a month.

DIAGNOSTIC CHARACTERS: The size and style of colouration (small spots rather than blotches) make the eggs of this Gull very distinct, and prevent their confusion with those of allied species breeding in our islands.

Family LARIDÆ. Genus LARUS.
Sub-family LARINÆ.

BLACK-HEADED GULL.

LARUS RIDIBUNDUS, *Linnæus.*

Single Brooded. Laying season, April, and prolonged into May.

BRITISH BREEDING AREA: The Black-headed Gull is one of the commonest and most widely distributed species of the present sub-family in the British Islands. From the nature of the country it is most abundant during the nesting season in Scotland and Ireland, but there are still several more or less important breeding-places in England. In the latter portion of our islands colonies of this bird are situated in Dorset, Kent, Essex, Suffolk, Norfolk, Lincolnshire, Staffordshire, Lancashire (with Walney), Yorkshire, Cumberland, and Northumberland. In Scotland there are many colonies of varying size (even as far north as the Shetlands) too numerous for enumeration, but special mention might be made of those near Glasgow and Loch Lomond. The same remarks apply to Ireland, where it must be stated the most westerly breeding range of the species obtains on the Blaskets, a small colony tenanting the island of Beginish.

BREEDING HABITS: The Black-headed Gull is a resident in the British Islands, but subject to much local and southern movement during the non-breeding season. Unlike most other British species in the present group, it frequents inland haunts, especially during the nesting season. Its favourite breeding-grounds are marshes and wet land, especially on islands, where greater immunity from danger is obtained. The presence of trees and bushes is by no means an obstacle, for many colonies are made on ground studded with low bushes and trees, or are completely surrounded by woods and plantations. This Gull is very gregarious during the breeding season, some of its colonies consisting of many thousands of pairs; everywhere the same social instinct is manifest, even though the assemblage numbers but a few individuals. In March the Gulls begin to congregate at the old colonies, so that we may infer they pair for life, many of these breeding-stations having been in use for time immemorial. A month later the nests are being made or repaired. These are built in the majority of instances on the ground, but odd pairs have been known to make them in trees, or even on boat-houses. The nests are placed in tufts of rushes, in hassocks of coarse grass, amongst reeds in shallow water, on heaps of dead, broken vegetation, or even on the flat ground, either covered with spongy moss or bare of all vegetation whatever. In many cases the nest is little more than a hollow in the ground or tuft, sometimes roughly lined with a little dry grass; in others it is better made, banked high above the surrounding marsh or shallow water, and made of reeds, rushes, flags, and coarse grass. As incubation advances, many of the nests made in the shallows are increased in bulk, so that the newly-hatched young may have plenty of accommodation on which to nest, and doubtless also for the purpose of providing against a sudden rise of the water, or the incessant wash of the tiny waves. When the colony is invaded, the Gulls rise in clouds from their nests, and commence a noisy clamour of remonstrance. The scene, once witnessed, can never be forgotten, and may be described best as a snowstorm, in which each flake is a fluttering, noisy Gull.

RANGE OF EGG COLOURATION AND MEASUREMENT: The eggs of the Black-headed Gull are usually three in number, but four are sometimes found. They vary considerably in colour, style of markings, shape, and size. They vary in ground colour from rich brown through every shade to pale bluish-green, spotted, blotched, and sometimes blurred or streaked with dark brown, and with underlying markings of violet-gray. On many varieties the markings are mostly distributed in an irregular zone round the larger end of the egg. Very eccentric-looking eggs may sometimes be found with the colouring matter in a circular patch on the larger end, gradually tinting off and fading into the ground colour round the margin; others may be seen with one or two large blotches or clouds of colour here and there, and the remainder of the shell free from markings. Average measurement, 2·2 inches in length, by 1·5 inch in breadth. Incubation, performed by both sexes, lasts from twenty-two to twenty-four days. This Gull will continue to lay clutch after clutch of eggs as they are removed; in many places the eggs are gathered for food, the poor birds patiently submitting to regular and systematic pillage every year, and yet continuing to haunt the old colony with a persistence that deserves a better reward.

DIAGNOSTIC CHARACTERS: The small size and characteristic Larine colouration distinguish the eggs of this Gull from those of all other allied species breeding in our islands.

Family LARIDÆ.   Genus STERNA.
Sub-family *STERNINÆ*.

SANDWICH TERN.

STERNA CANTIACA, *Gmelin*.

Single Brooded. Laying season, middle of May to middle of June.

BRITISH BREEDING AREA: The Sandwich Tern has been exterminated as a breeding species from many localities, partly by incessant egg collecting, and partly by the rise of tourist resorts on once little frequented parts of the coast where it nested. The most important breeding-place in our islands is on the Farne Islands, but even here it is much less abundant than formerly. Twenty years ago I well remember bushels of its eggs were sent from these islands into Sheffield for sale! Small colonies still exist on Walney Island, off the Lancashire coast, and at Ravenglass in Cumberland; a few pairs breed in the Solway district, on Loch Lomond, in the Firth of Tay, on the coast of Elgin, and perhaps in Sutherland. In Ireland it is known to breed sparingly in Co. Mayo, on an island in a moor loch some miles from the sea.

BREEDING HABITS: The Sandwich Tern is a summer migrant to the British Islands, reaching our coasts during April and May. The favourite breeding-haunts of this beautiful Tern are low rocky or sandy islands, covered with a good growth of campion, thrift, and long grass, but varied here and there with patches of bare ground, and with a beach of rough shingle. Similar conditions are chosen on the mainland in some secluded part of the coast, but everywhere an island is preferred. The birds doubtless pair for life, and continue to visit certain places for the purpose of nesting every season. This Tern is gregarious, but the colonies vary a good deal in extent, that on the Farne Islands being by far the most important and time-honoured. It will be remarked, however, that the same patch of ground is not invariably used, the birds selecting certain sites in succession. But little nest is made, merely a hollow lined with a few bits of withered marine herbage; whilst in many cases a nest of any kind is dispensed with altogether. Many nests are placed near to each other, in some cases not more than a foot apart. These are made either amongst the sand, shingle, and drift near the water, amongst short grass and campion, and more frequently further away from the sea, on a bare patch of elevated ground. The birds rise in clouds as soon as their haunt is invaded, and fluttering and screaming hover above the intruder's head as he walks amongst the nests. The eggs resemble the surroundings so closely in tint that great care is needed in walking not to tread upon them.

RANGE OF EGG COLOURATION AND MEASUREMENT: The eggs of the Sandwich Tern are usually two in number, more rarely three, and vary considerably in colour. In ground colour they vary from white or creamy-white to buff, spotted and blotched with dark brown and orange-brown, and with underlying markings of violet-gray. It is impossible adequately to describe these exceedingly handsome and richly-marked eggs. Some varieties are covered with large bold blotches and washes of colour, others are splashed here and there with brown or gray; some are evenly spotted over the entire surface, others zoned, others covered with short, streaky lines. Average measurement, 2·1 inches in length, by 1·4 inch in breadth. Incubation, performed by both sexes, lasts from twenty-one to twenty-four days. If the first lot of eggs be taken or washed away by high tides, as sometimes happens, others will be laid.

DIAGNOSTIC CHARACTERS: The size and remarkably rich colouration readily distinguish the eggs of the Sandwich Tern from those of allied species breeding in our islands.

Family LARIDÆ.   Genus STERNA.
Sub-family STERNINÆ.

ROSEATE TERN.

STERNA DOUGALLI, *Montagu.*

Single Brooded.   Laying season, latter end of May, and in June.

BRITISH BREEDING AREA: It is with some hesitation that I include the Roseate Tern in the present work. It is sad to relate that a bird actually first made known to science from examples obtained in Scotland, and described by Montagu, has become almost if not quite extinct as a breeding species in our islands. It formerly bred on the Scilly Islands, on Foulney and Walney Islands, the Farne Islands, on the Cumbrae Isles in the Firth of Clyde, and on several other islets off the Scotch and Irish coasts. Pairs have been identified during recent years on the Farne Islands, but no evidence of their breeding is forthcoming. In the hope that this beautiful bird will re-establish itself in its former haunts at no remote date (now that our sea birds are fortunately protected by law), I devote a short chapter to its nesting economy.

BREEDING HABITS: The Roseate Tern is (or was) one of the latest of our summer migrants, not reaching its breeding-stations until towards the end of May. Its favourite haunts are low, rocky islands, especially such that contain reaches of sand and shingle; in this respect, as in fact in most of its reproductive economy, it very closely resembles its near ally, the Common Tern. It is not known whether this Tern pairs for life or not, but probably such is the case. It does not appear to make any nest, but lays its eggs on the bare ground, often in a slight hollow amongst the shingle. Its behaviour at the breeding-grounds, when disturbed, is not known to differ from that of allied species.

RANGE OF EGG COLOURATION AND MEASUREMENT: The eggs of the Roseate Tern are two or three in number, and so closely resemble those of the Common and Arctic Terns that a detailed description is unnecessary. The range of colouration is precisely similar. Average measurement, 1·7 inch in length, by 1·15 inch in breadth. The period of incubation is not known to differ from that of the Common or Arctic Terns.

DIAGNOSTIC CHARACTERS: It is impossible to distinguish the eggs of the Roseate Tern from those of the two closely allied species breeding in our area. No eggs require more careful identification, and if other Terns are breeding in the vicinity, the only safe way is to watch the parent birds settle upon their eggs before taking them. The adult birds may be distinguished by their roseate under parts, and by the white margins to the inner webs of the primaries.

Family LARIDÆ.   Genus STERNA.
Sub-family *STERNINÆ*.

COMMON TERN.

STERNA HIRUNDO, *Linnæus*.

Single Brooded.   Laying season, end of May and June.

BRITISH BREEDING AREA: The Common Tern is an abundant and widely distributed species, but becomes rarer in the northern portions of our area, where it is replaced to a more or less extent by the Arctic Tern. It has breeding-places here and there in suitable districts round the coasts of England and Wales. Scattered colonies also occur on both the east and west coasts of Scotland, at least as far north as Skye on the latter, and Moray Firth (in nearly the same latitude), the east coast of Sutherlandshire to the Pentland Skerries, and perhaps the Orkneys, on the former. In Ireland it is more abundant than the Arctic Tern, breeding here and there round the coast, and also on the banks of some of the inland lakes.

BREEDING HABITS: The Common Tern is a summer migrant to our islands, arriving towards the end of April or early in May. Its favourite haunts are low rocky islands covered with marine vegetation, links, and shingly reaches on quiet, secluded parts of the coast. It is a gregarious species, but the colonies vary a good deal in size, owing to local causes. One of the most extensive colonies is situated on the Farne Islands. This Tern probably pairs for life, and yearly returns to one particular spot to breed, to which the birds are greatly attached. The nests are invariably made on the ground, amongst sea-campion, thrift, or grass, or on some spot where the earth is bare of herbage. I do not think as a rule, that this Tern nests so near the water as the Arctic Tern, and shows more preference for laying amongst or near vegetation of some kind. The nest is a mere hollow, scantily lined with bits of dry grass and withered marine vegetation. The behaviour of this Tern at the nest is precisely similar to that of its congeners. As soon as the breeding-ground is invaded the birds rise in a noisy, anxious, fluttering throng, and continue flying restlessly about until their haunt is left in peace.

RANGE OF EGG COLOURATION AND MEASUREMENT: The eggs of the Common Tern are two or three in number, and subject to some considerable amount of variation. They vary from buff to grayish-brown in ground colour, spotted and blotched with dark brown and yellowish-brown, and with underlying markings of gray. On some eggs the blotches are large, few, and irregular; on others they take the form of spots, and are often distributed in a zone round the larger end; occasionally a few streaks occur, and some varieties are evenly marked over most of the surface with small spots. Average measurement, 1·7 inch in length, by 1·2 inch in breadth. Incubation, performed by both sexes, lasts from twenty-one to twenty-three days.

DIAGNOSTIC CHARACTERS: The eggs of this Tern very closely resemble those of the Arctic Tern, but may almost invariably be distinguished by their larger bulk, more rotund form, and absence of any olive or green tinge in the ground colour. They should, however, be carefully identified wherever possible.

Family LARIDÆ.　　　　　　　　　　　　　Genus STERNA.
Sub-family STERNINÆ.

ARCTIC TERN.

STERNA ARCTICA, *Temminck*.

Single Brooded.　Laying season, June.

BRITISH BREEDING AREA: The Arctic Tern is a common and widely distributed species, especially in the northern portions of our area, where it replaces the preceding bird in many localities, or far exceeds it in numbers. It breeds locally and in most cases sparingly round the English coast-line, from the Farne Islands to the sand-banks near the Spurn. No other stations are known until we reach the Scilly Islands; thence northwards a few pairs breed on the Welsh coast, and possibly on Walney, off the Lancashire coast. In Scotland it is generally distributed round the coast in places suited to its requirements, including the Orkneys, the Shetlands, and the Hebrides. Its principal breeding-places in Ireland are on the west coast.

BREEDING HABITS: The Arctic Tern is a summer migrant to the British Islands, arriving on the coasts towards the end of April or early in May. Its favourite breeding-haunts are low rocky islands where the beach is covered with sand and shingle, and the ground is more or less clothed with grass and other marine vegetation. In some localities, where an island cannot be had, a secluded part of a shingly beach is selected. Like all its congeners, the Arctic Tern is gregarious during the breeding season, but the colonies vary a good deal in size. One of the most important of these is on the Farne Islands. This Tern probably pairs for life, and yearly returns to the same islands to breed, although the exact locality of the colony is frequently changed from season to season. From a very long and wide experience of the habits of this Tern, I am of opinion that the bird never makes a nest of any kind, beyond a mere hollow, which is as often due to accident as to design. No lining is ever used. The eggs also are more generally laid nearer to the water than those of the Common Tern. Any little depression in the coarse sand or shingle, on the line of drift, among pebbles, or on the bare ground or rock, serves as a receptacle for the eggs. The actions of this Tern at the nest are precisely similar to those of the preceding species. The eggs are seen with difficulty, owing to their harmonizing with surrounding tints.

RANGE OF EGG COLOURATION AND MEASUREMENT: The eggs of the Arctic Tern are two or three in number, and vary considerably in colour. They vary in ground colour from buff or buffish-brown to olive, or even pale bluish-green, blotched and spotted with dark brown and ochraceous brown, and with underlying markings of gray. The distribution of the markings runs through much the same variation as those of the preceding species, but blotches are more general. Average measurement, 1·5 inch in length, by 1·1 inch in breadth. Incubation, performed by both sexes, lasts from twenty-one to twenty-three days.

DIAGNOSTIC CHARACTERS: The eggs of the Arctic Tern are distinguished from those of the Common Tern, the only species with which they are likely to be confused, by their more elongated shape, smaller size, bolder markings, and proneness to olive tints. Care, however, should be exercised in identifying them

Family LARIDÆ.   Genus STERNA.
Sub-family STERNINÆ.

LESSER TERN.

STERNA MINUTA, *Linnæus*.

Single Brooded.   Laying season, June.

BRITISH BREEDING AREA: The Lesser Tern is fast becoming local and even rare, partly owing to the ruthless plundering of its eggs by sea-side "trippers," and persecution by cockney sportsmen, and partly because many of its favourite haunts or their vicinity have now become too thickly populated by man. I will not assist in this Tern's further extermination by naming a single haunt where I know it now breeds, beyond remarking in a vague and general way that small colonies exist here and there round the British coasts, and in one or two more inland localities. When we secure by law more rigorous protection for this, my favourite sea-bird, then may its haunts be more definitely named.

BREEDING HABITS: This charming little Tern is a summer migrant to our islands, rarely reaching them before May. Its haunts during the breeding season are wide stretches of sandy coast, varied with slips and banks of shingle. Curiously enough it prefers the coast of the mainland to an island, and this undoubtedly is one of the principal causes of its rapid decrease in numbers, owing to the facilities offered to every wandering rascal to plunder and destroy. Like all the other Terns, the present species probably pairs for life, and yearly returns to one particular part of the coast to breed. It is also gregarious, like its congeners, but unfortunately its colonies in our islands are nowhere very extensive. This Tern makes no nest, not even scratching a hollow for the reception of the eggs, but laying them, not on the fine sand, but on the slips of rough shingle, where the bold character of their markings harmonizes so closely with the ground that their discovery is difficult. The birds become noisy and fly anxiously about overhead when their nesting-ground is invaded, but show less concern for their treasures than the other Terns invariably do.

RANGE OF EGG COLOURATION AND MEASUREMENT: The eggs of the Lesser Tern are from two to four in number. On two separate occasions I have seen clutches of the latter, but three is the average number. They vary from buff to grayish-brown in ground colour, spotted and blotched with dark brown and yellowish-brown, and with underlying markings of gray. They run through every variation that characterizes the eggs of the two preceding species, and may perhaps be best described as resembling those of the Common Tern in ground colour, and those of the Arctic Tern in the bold nature of the markings. Average measurement, 1·25 inch in length, by ·95 inch in breadth. Incubation, performed by both sexes, is said by Tiedemann to last from fourteen to sixteen days, but this probably is an error, the period most likely being the same as that of the preceding species.

DIAGNOSTIC CHARACTERS: The eggs of the Lesser Tern may be at once distinguished by their small size, elliptical shape, and the bold character of their markings.

Family ALCIDÆ.   Genus URIA.

COMMON GUILLEMOT.

URIA TROILE (*Linnæus*).

Single Brooded.   Laying season, May and June.

BRITISH BREEDING AREA: The Common Guillemot is generally and abundantly distributed along most of the rocky coasts of the British Islands, from the Scilly Isles and the Isle of Wight in the south, to the Orkneys and Shetlands in the north; from the Flamborough cliffs and the Farne Islands in the east, to the Outer Hebrides, St. Kilda, and the Blaskets in the west. It is, however, as far as my experience goes, much more abundant in the north than in the south, and frequents sandstone cliffs from necessity rather than choice.

BREEDING HABITS: The Guillemot, most familiar of all our rock fowl, is a resident in our islands, but subject to much local movement when the breeding season is over. It is gregarious during the nesting season, some of its colonies being densely packed. Where the cliffs are extensive the Guillemots are more scattered, but where suitable crags are few the birds mass together in enormous numbers, as, for instance, at the Pinnacles on the Farne Islands. Cliffs with plenty of ledges and hollows are the favourite nesting-places. The bird may doubtless pair for life, and certain haunts are returned to year by year. The Guillemot is no nest-builder, nor does it make provision in any way whatever for the reception of its egg. This is deposited on any suitable ledge in any available hollow, where it can be tolerably safe from rolling over, or the parent bird can find standing room to incubate it. The Guillemot is not very demonstrative at the breeding-place, usually flying down

PLATE X.

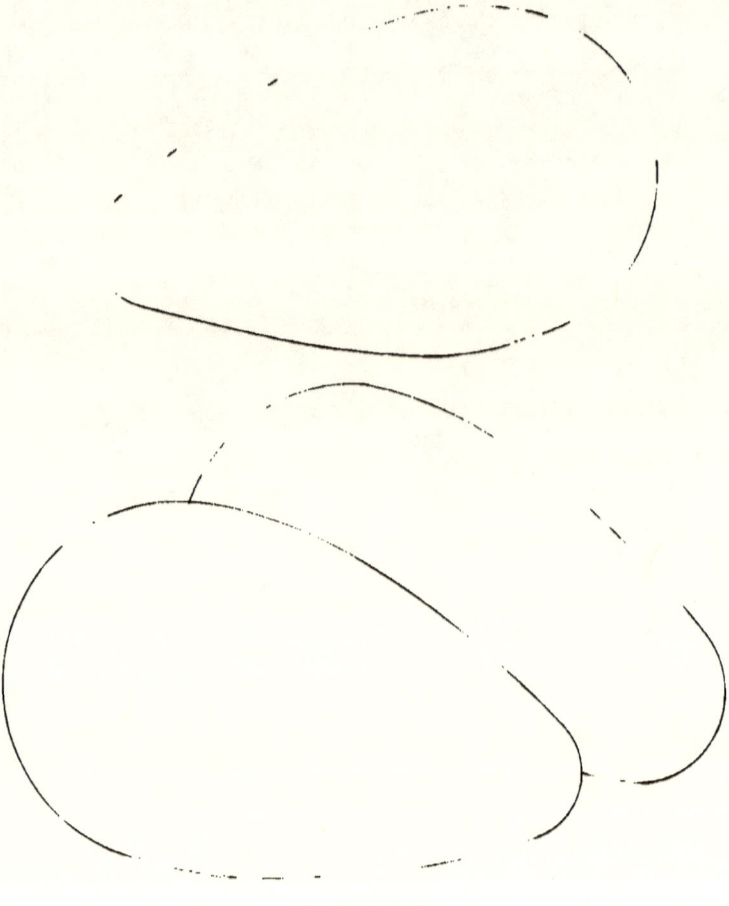

GUILLEMOT

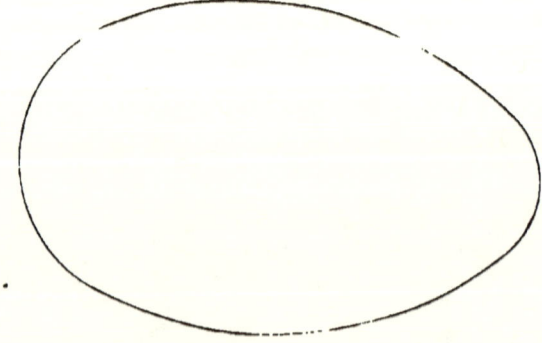

RED-THROATED DIVER

to the sea when disturbed, and returning as soon as its haunt is left in peace. It utters little or no sound when thus disturbed.

RANGE OF EGG COLOURATION AND MEASUREMENT: The Guillemot lays only one egg, but if this be taken it is usually replaced several times in succession. It is impossible, even with much more space than we here have at our disposal, to describe a tithe of the variations in colour and character of markings in the eggs of this species. In no other known bird do they vary so widely; and certainly the eggs of none are more beautiful. The ground colours of the egg (which is pear-shaped) are dark green, yellowish-green, reddish-brown, cream-yellow, white, and pale blue, with every intermediate shade; the markings, which take the form of spots, blotches, streaks, and zones, are composed of browns and grays and pinks of every possible tint. Some eggs of course are much handsomer than others; some are without markings of any kind. An exquisite variety is white, intricately streaked and netted with pink; another is green, streaked in the same manner with yellow, light brown or nearly black; others are zoned with blotches, or marked with fantastic-shaped spots and rings. Average measurement, 3·3 inches in length, by 2·0 inches in breadth. Incubation, performed by both sexes, lasts from thirty to thirty-three days.

DIAGNOSTIC CHARACTERS: The egg of the Guillemot, in spite of its wonderful variation, cannot readily be confused with that of any other species, except that of the Razorbill. Eggs, however, that resemble those of the latter bird may be at once distinguished by holding them up to the light and looking at the interior of the shell through the hole where they have been blown, when the colour is yellowish-white, never green. See also diagnostic characters of the egg of the Razorbill (p. 315).

Family ALCIDÆ. Genus URIA.

BLACK GUILLEMOT.

URIA GRYLLE (*Linnæus*).

Single Brooded. Laying season, latter half of May and in June.

BRITISH BREEDING AREA: The Black Guillemot is only known to breed in one English locality, and that is the Isle of Man. Its principal breeding area extends along the west and north coasts of Scotland, including the Hebrides and St. Kilda, to the Orkneys and the Shetlands. Thence it extends down the east coast of Scotland as far south as Sutherlandshire. Its principal breeding-places in Ireland are on the north and west coasts, but it is said there are still a few on the east coast and on the south coast as far as County Waterford.

BREEDING HABITS: The Black Guillemot is found in the British seas throughout the year, but is subject to considerable local movement after the breeding season. Its haunts in our islands during the summer are bold rocky headlands, islands, and ocean cliffs similar to those frequented by the Razorbill and the Common Guillemot. It is not quite so gregarious as its larger allies, the colonies usually being small and scattered. I am of opinion that this species pairs for life, as year by year certain spots are resorted to for breeding purposes. Unlike the Guillemot, but like the Razorbill, the present bird breeds in a covered site, in holes and crannies of the cliffs, or amongst masses of rocks below them, occasionally on rock-strewn downs sloping to the sea. No nest of any kind is made, the eggs simply resting on the bare earth or rock. This bird makes little or no demonstration when disturbed,

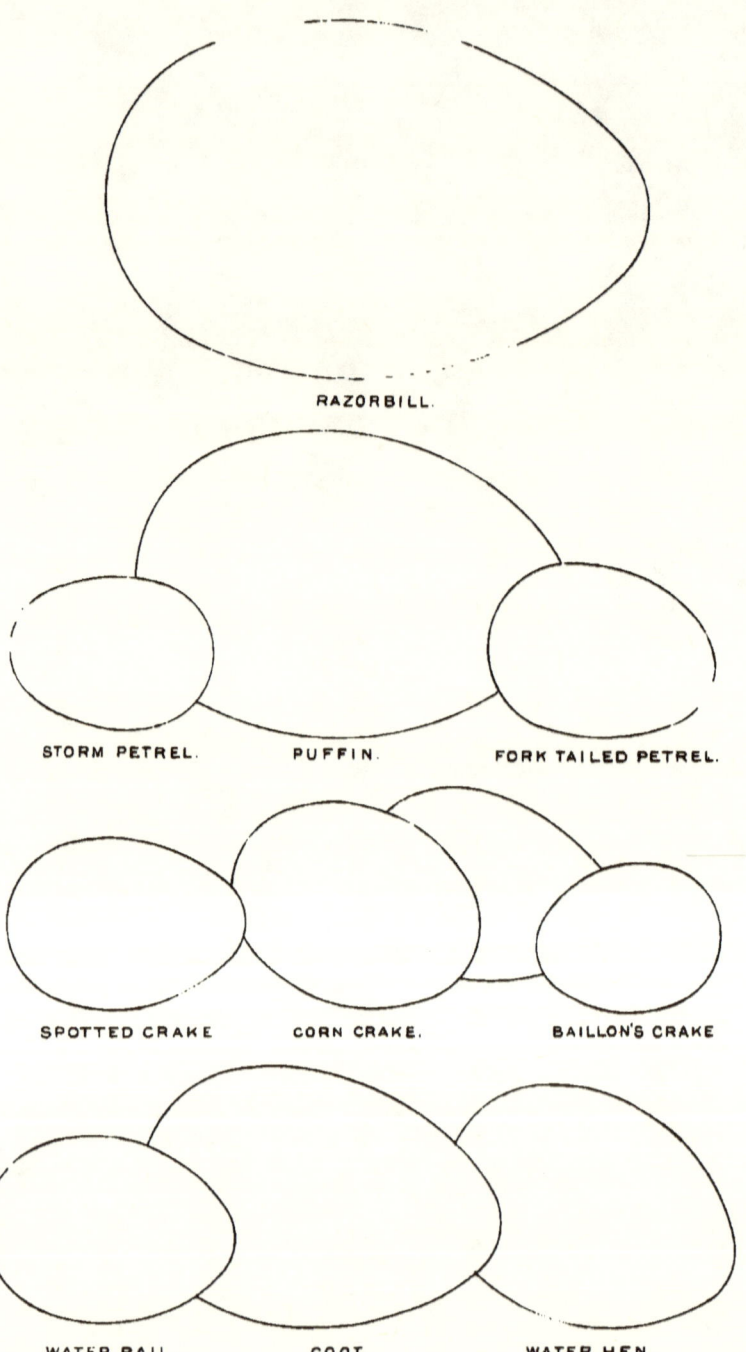

usually flying off at once to the water, there to remain until the danger has passed.

RANGE OF EGG COLOURATION AND MEASUREMENT: The eggs of the Black Guillemot are two in number, smaller and blunter than those of the Common Guillemot, and resembling those of the Razorbill in general colouration. They vary from cream to buff and pale green in ground colour, blotched, spotted, and more rarely streaked with rich dark brown and paler brown, and with numerous and large underlying markings of gray. Usually the blotches and spots are pretty evenly distributed over the entire surface, but zones are not uncommon. Average measurement, 2·35 inches in length, by 1·6 inch in breadth. Incubation, performed by both sexes, is said to last twenty-four days.

DIAGNOSTIC CHARACTERS: The eggs of the Black Guillemot may be readily distinguished from those of allied species by their much smaller size. They resemble, however, certain varieties of those of the Sandwich Tern, but may be at once distinguished by their greater weight and greenish tinge. The site on which they are laid also prevents the slightest possible chance of confusion.

Family ALCIDÆ. Genus ALCA.

RAZORBILL.

ALCA TORDA, *Linnæus*.

Single Brooded.   Laying season, middle of May to middle of June.

BRITISH BREEDING AREA: The Razorbill is widely and commonly distributed along all rocky coasts suited to its requirements, although nowhere perhaps so abundant as the Guillemot, and becoming less frequent in the extreme south.

BREEDING HABITS: The Razorbill is a resident in the British seas, but subject to much local movement during the non-breeding season. The breeding-haunts of this species are marine crags and precipices, in which plenty of nooks and crannies and broken rocks occur. The broken cliffs are preferred to those of a wall-like character. This bird is also gregarious, and great numbers may be found breeding together in suitable cliffs. I am of opinion that the Razorbill pairs for life, as each season one particular hole or cranny will be tenanted, presumably by the same birds. Early in spring the birds begin to collect at the nesting colonies, coming from all parts of the adjoining seas to the old familiar rendezvous. This species makes no nest of any kind, laying its egg in a crevice or hole in the cliffs, or far under a stack of rocks poised one on the other, where to reach it is absolutely impossible. I have known this bird make use of a Puffin burrow; but surely the ornithologist who declares that he saw a Razorbill incubating in the old nest of a Cormorant must have been deceived! The Razorbill is somewhat loth to leave its egg, and may often be caught on it owing to this reluctance. The birds also occasionally make a grunting noise when their breeding-grounds are invaded by man.

RANGE OF EGG COLOURATION AND MEASUREMENT: The Razorbill lays only a single egg, which will, however, be several times replaced if taken. The eggs of this species vary considerably in colour and character of markings, but to nothing near the extent of those of the Guillemot. The ground colour runs through every tint between white and reddish-brown, and the blotches and spots are dark liver-brown and reddish-brown, whilst the underlying markings are gray or grayish-brown. The markings are usually very bold and large, most numerous about the larger end of the egg, where they frequently form an irregular zone. One variety has the markings of a streaky character; another and rarer variety is almost spotless. No shade of green or blue is ever displayed externally on the eggs of this species. Average measurement, 2·9 inches in length, by 1·9 inch in breadth. Incubation, performed by both sexes, lasts from thirty to thirty-three days.

DIAGNOSTIC CHARACTERS: The eggs of the Razorbill are never so acutely pear-shaped as those of the Guillemot, being blunter and broader for their length. They can only possibly be confused with those of the Guillemot, but from which they may be readily distinguished by the clear pea-green tinge of the interior of the shell when held up to the light; those of the Guillemot always being yellowish-white, except in such varieties where the green or blue exterior overpowers it. As previously remarked, however, the eggs of the Razorbill externally never exhibit green or blue tints.

Family ALCIDÆ.　　　　　　　　　　Genus FRATERCULA.

PUFFIN.

FRATERCULA ARCTICA (*Linnæus*).

Single Brooded.　Laying season, May and early June.

BRITISH BREEDING AREA: Except on the south of England, where it is only sparingly dispersed, and the east of England, where it is not known to breed south of Flamborough, the Puffin is widely and generally distributed round the British Islands, in some places, as for instance Lundy Island, the Farne Islands, and the Hebrides, its numbers being enormous and past all belief.

BREEDING HABITS: The Puffin is a resident in the British seas, but subject to much local movement after the breeding season. Its favourite breeding-haunts are the earthy parts of sea-cliffs, sloping downs covered with turf, and low islands with a good depth of soil or peat. At the Bass Rock many Puffins have formed a colony amongst the walls of an old fortress; at St. Kilda I noticed another considerable colony in the sandy cliff near the village, whilst the cliff Connacher and the island of Doon at the latter place (St. Kilda) contain colonies of this bird that perfectly overpower one with their countless numbers. Few birds are so gregarious as the Puffin. It also probably pairs for life, and the same places are resorted to annually. In most localities the Puffin makes its nest at the end of a long and often winding burrow, which is usually excavated by the birds themselves, or sometimes a natural hollow or crevice in the cliff or under a mass of fallen rocks is utilized. More rarely a rabbit-burrow is annexed. By the end of April both birds are engaged in making this burrow, if circumstances demand it, which is often several yards in length, though more often only a few feet. At the end a slight nest of dry grass and sometimes a few feathers is formed. I have known several pairs to nest in one large earth. When disturbed at the breeding-places the Puffins that may chance to be outside their holes take wing, usually going at once to the sea with whirring wings, but not uttering a single note. Those on their nests, however, rarely move until they are pulled out. Great caution is required, and gloves are advisable, as the disturbed Puffin resents intrusion, and is able to inflict a nasty cut with its sharp beak and claws. The throng of Puffins I disturbed from the colonies at St. Kilda was perfectly overwhelming, and can never be forgotten.

RANGE OF EGG COLOURATION AND MEASUREMENT: The Puffin lays only one egg, which will, however, be replaced several times if taken. It is dull white, sometimes tinged with very pale blue or gray, in ground colour, very obscurely spotted, blotched, and occasionally streaked with pale brown, and with underlying markings of gray. As a rule the latter type of markings prevails, and very often both kinds form a zone round the larger end of the egg. So dirty, however, do the eggs become through contact with the wet feet and plumage of the parent birds and the soil of the burrow, that little, if any, of the pale markings are visible until they have been well washed. Average measurement, 2·4 inches in length, by 1·65 inch in breadth. Incubation, performed by both sexes, lasts about five weeks.

DIAGNOSTIC CHARACTERS: The egg of the Puffin is readily identified by its size, pale ground colour, and very indistinctly defined markings.

Family PROCELLARIIDÆ.   Genus PROCELLARIA.

LEACH'S FORK-TAILED PETREL.

PROCELLARIA LEACHI, *Temminck.*

Single Brooded.   Laying season, June.

BRITISH BREEDING AREA: The known breeding-places of the Fork-tailed Petrel within our area are remarkably few, but there can be no doubt whatever that many more will eventually be discovered on the western coast-line, notably amongst the Hebrides. It breeds abundantly at St. Kilda ; there are colonies on North Rona and elsewhere in the Outer Hebrides; whilst small colonies have within the past few years been discovered on the Blaskets off the coast of County Kerry.

BREEDING HABITS: The Fork-tailed Petrel is a resident in the British Seas, but subject to much local movement, and is seldom seen on or near land except during the breeding season, or when driven in by stress of weather. Its breeding-haunts, so far as the United Kingdom is concerned, are situated on rocky islands commanding the Atlantic Ocean. It is a gregarious bird, but nowhere do the colonies appear to be very extensive, being scattered here and there on the fringe of its oceanic haunts. Most probably this Petrel pairs for life, as yearly it may be found breeding in one particular spot, the same nesting sites being used in many cases. The most important colony known to me is situated at St. Kilda. The Fork-tailed Petrel makes its scanty nest at the end of a burrow from two to five feet or more in length, often very winding, and containing several outlets. In some cases one earth will accommodate several pairs of birds. These burrows are made in the soft peaty soil or mould, usually under the turf, near the summit of the cliffs ; but in some instances they have been discovered amongst ruins or under rocks. The nests I examined were made of dry grass, moss, roots, and a few bits of lichen from the surrounding rocks, and varied considerably in size: in one burrow there was no nest, and the egg lay on the bare ground. I also remarked that the inhabited burrows had a little dry grass strewn at the entrance. The birds sit remarkably close. No one would dream that the ground around him contained such interesting objects, the Petrels remaining on their nests until dragged out by the hand

Many nests may be found within the radius of a few yards.

RANGE OF EGG COLOURATION AND MEASUREMENT: The Fork-tailed Petrel only lays one egg each season. This is chalky in texture, very fragile, white in ground colour, with a more or less obscure zone of minute reddish-brown specks, and a few underlying ones of grayer brown. So fine are most of these markings that they look almost like dust. Average measurement, 1·3 inch in length, by ·97 inch in breadth. Incubation, performed by both sexes, lasts about thirty-five days.

DIAGNOSTIC CHARACTERS: The eggs of this Petrel are readily distinguished from those of all other allied species breeding in our area by their size.

Family PROCELLARIIDÆ.   Genus PROCELLARIA.

STORMY PETREL.

PROCELLARIA PELAGICA, *Linnæus*.

Single Brooded.   Laying season, June (normally).

BRITISH BREEDING AREA: So far as is known the Stormy Petrel does not breed anywhere on the eastern coast-line of England or Scotland. It breeds on the Scilly Isles, and is said to do so sparingly on Lundy; a few nest here and there on the Welsh coast, but it becomes more generally distributed along the rugged and islet-studded coast of the west of Scotland, including the Hebrides and St. Kilda, and round the north coast to the Orkneys and the Shetlands. There are also numerous breeding-places round the Irish coasts, especially in the west, including the Blaskets.

BREEDING HABITS: The Stormy Petrel is a resident in the British seas, subject to much local movement, and only coming to land to breed, or when driven in by stress of weather. Its favourite breeding-haunts are rocky islands which contain a fair amount of uneven turf-clad downs more or less strewn with rock fragments. It is gregarious during the breeding season, numbers of pairs resorting to certain favourite localities to rear their young. It is probable that this little Petrel pairs for life. The slight nest of a few bits of dry grass is made either in an old rabbit-earth or Puffin burrow, under a rock or a heap of loose stones, or even amongst ruined walls and other masonry. In some cases no provision of any kind whatever is made, and the egg rests on the bare ground. This Petrel is a remarkably close sitter, so that one could wander about its breeding-places without being made aware of the fact. It sits close in its burrow or crevice until dragged out; at dusk, however, the birds become more lively, and may then be seen flitting about in a ghostly way near their nest-holes, leaving and returning to them. Many nests are often made close together.

RANGE OF EGG COLOURATION AND MEASUREMENT: The Stormy Petrel only lays one egg each season, but if this chance to be taken it is generally replaced. It is pure white in ground colour, chalky in texture and without gloss, and almost oval in form. From a casual glance many eggs look entirely spotless, but a fine sprinkling of minute red spots, most of them in the form of a zone round the larger end, in somes cases so small as to look like dust, may be detected. Average measurement, 1·1 inch in length, by ·83 inch in breadth. Incubation, performed by both sexes, lasts from thirty-three to thirty-five days.

DIAGNOSTIC CHARACTERS: The small size, chalky

texture, and minute, dust-like markings readily distinguish the egg of the Stormy Petrel from those of all other species breeding in our area with which it is at all likely to be confused.

Family PROCELLARIIDÆ.　　　　　　　Genus FULMARUS.

FULMAR PETREL.

FULMARUS GLACIALIS (*Linnæus*).

Single Brooded. Laying season, middle of May to middle of June.

BRITISH BREEDING AREA: The only important British breeding-place of this Petrel is in the St. Kilda group. It is said, on what appears to be trustworthy authority, that a few pairs breed on Foula in the Shetlands, but I cannot accept the evidence of its doing so on the Flannans and the Seven Hunters off the coast of Lewis until less ambiguous and playing-for-safety kind of statements are made than "recent evidence points to the establishment of colonies on" those islets, as a modern master was condescending enough to inform his befogged and bewildered readers!

BREEDING HABITS: The Fulmar is more or less a resident in the British seas, but subject to much nomadic movement when the breeding season is over. Its breeding-haunt in our area is the lofty stacks and beetling precipices of St. Kilda and the neighbouring isles—the vast downs and crags and turf-grown cliffs that spring from the wild open Atlantic to a culminating height of twelve hundred feet above the waves. No other part of our sea-girt islands is more grandly majestic than the Fulmar's haunt. The bird is eminently gregarious during the breeding season, and here clusters in tens of thousands to rear its young each year. The Fulmar probably pairs for life; it is much attached to its breeding-place, and the nests are used year by year. The nest is made on the face of the cliff, either on rough ledges or in crevices and hollows amongst the piled-up rock masses. The favourite situation is on that portion of the cliff where a good layer of turf-clad soil is present; the bird evidently preferring to burrow a short distance into the ground wherever possible. The hole, however, is seldom big enough entirely to conceal the bird, and in most cases does not more than half conceal it; whilst in a great many cases it shelters itself under a projection of earth and turf. The nest is slight, merely a little dry grass, and in many cases even this is dispensed with. I met with a peculiar type of nest on some of the bare rock-ledges and in crevices, consisting entirely of small bits of rock arranged very neatly. Vast numbers of birds nest close together, so near in many parts of the cliffs as to almost touch each other, and looking at a distance like masses of snow. The birds are quite silent when disturbed, but the impressive scene of the fluttering, drifting, feathered hosts is beyond all description. As a rule the Fulmars are somewhat loth to leave their nests, and the natives take advantage of this, and snare thousands as they sit on their egg.

RANGE OF EGG COLOURATION AND MEASUREMENT: The Fulmar only lays a single egg, and it is said that if this be taken no more are produced that season. It is white and spotless, rough and chalky in texture, and with a strong peculiar pungent smell. Average measurement, 2·9 inches in length, by 2·0 inches in breadth.

Incubation, performed by both sexes, lasts from fifty to sixty days according to Thienemann.

DIAGNOSTIC CHARACTERS: The egg of the Fulmar is readily distinguished by its size, absence of markings, and pungent smell.

Family PROCELLARIID.E. Genus PUFFINUS.

MANX SHEARWATER.

PUFFINUS ANGLORUM (*Temminck*).

Single Brooded. Laying season, May and first half of June.

BRITISH BREEDING AREA: The Manx Shearwater has no known breeding-place on the east coast of Scotland, or the east and south coasts of England. It breeds, however, on the Scilly Isles, possibly on Lundy, here and there on the Welsh coast, and in many parts of the west of Scotland among the Hebrides, including St. Kilda, and northwards to the Orkney and Shetland groups. In Ireland it breeds on some parts of the coasts of Waterford and Wexford, possibly on the Blaskets, certainly on the Skelligs rocks, various islets off the coast of Donegal, and on Rathlin. Many other stations undoubtedly yet remain to be discovered.

BREEDING HABITS: The Manx Shearwater is a resident in the British seas, but during the non-breeding season is subject to much nomadic movement. Its favourite breeding-places are islands with a good ocean aspect, especially such as are broken into grassy downs, and fall in crags and precipices more or less turf-grown. The bird is especially fond of a wide sloping down gently falling to a rocky beach on one side, and on the other culminating in rugged precipices full of turfy hollows with a thick covering of loamy soil. It is gregarious during the breeding season, but the colonies vary a good deal in size, the most important and extensive one known to me being on the island of Soay in the St. Kilda group. Many scattered pairs may be met with here and there, but this island seems to be the grand head-quarters of the species in our area, if we perhaps except that on Puffin Island near the Little Skellig. This Shearwater most probably pairs for life, and season by season returns to its favourite breeding-places. The egg is laid in a burrow, usually excavated by the bird, and from four to twelve feet or more in length. These burrows are made in the steep grassy parts of the cliff, or near their summit, or on the downs sloping to the water. Many of them are made under large masses of rocks impossible of human access. I found a scanty nest of dry grass at the end of the burrow, but other observers have found the egg on the bare ground. At the entrance of a burrow in use a few droppings are almost invariably seen. The Manx Shearwater sits closely; in fact, few if any of the birds are to be seen during daylight, no matter how large the colony may be. The sitting bird rarely makes any effort to escape, suffering itself to be dragged out. Many nests are made close together, and sometimes one main entrance will lead to several burrows each containing a nest.

RANGE OF EGG COLOURATION AND MEASUREMENT: The Manx Shearwater only produces a single egg for a sitting, but if this be removed another will generally be laid. This is pure white in colour, and rather smooth in texture. Average measurement, 2·4 inches in length, by 1·65 inch in breadth. Incubation is performed by both sexes, but the period of its duration is not apparently ascertained.

DIAGNOSTIC CHARACTERS: The egg of the Manx Shearwater is readily distinguished from all other Petrels known to breed within our area by its size. Care, however, should be taken in its identification, as it is by no means improbable that other closely allied Shearwaters may breed on the British coasts, although hitherto undetected.

Family COLYMBIDÆ.   Genus COLYMBUS.

RED-THROATED DIVER.

COLYMBUS SEPTENTRIONALIS, *Linnæus*.

Single Brooded. Laying season, latter half of May, and in June.

BRITISH BREEDING AREA: The Red-throated Diver is principally confined to the western half of Scotland from the Clyde northwards, including the Hebrides, the Orkneys, and the Shetlands. In Ireland it is at present only known to breed in Co. Donegal. The bird is only a winter visitor to the English coasts.

BREEDING HABITS: The Red-throated Diver is a resident in the British Islands, but subject to much local and southern movement after the breeding season. The favourite breeding-haunts of this Diver are the wild moorland lochs and pools that form so characteristic a feature in the scenery of the Highlands and the Hebrides. It is not gregarious during the nesting season, although several pairs may frequent one loch or pool, each, however, keeping to a particular locality and resenting any intrusion. The bird most probably pairs for life, and returns season after season to breed in certain chosen spots. The nest is invariably placed on the ground, the bird choosing an island wherever a preference can be exercised; but frequently the bank of a mountain pool is selected. If the nest is made on shingle or bare ground it is more elaborate than when on grass or amongst vegetation. Even the most elaborate nests are little more than hollows sparsely lined with a little dry grass or fragments of more aquatic herbage. The nest is seldom far from the water-side. The bird does not sit very closely, usually gliding off the nest and shuffling into the water the moment it is alarmed. Neither is it a demonstrative bird, but it sometimes flies over the intruder in its anxiety to learn the fate of its home.

RANGE OF EGG COLOURATION AND MEASUREMENT: The eggs of the Red-throated Diver are two in number, narrow and elongated in form. They vary from brownish-olive to pale buffish-brown in ground colour, spotted and speckled with very dark brown, many of them almost black, and with a few indistinct underlying markings of paler brown. The spots vary from about the size of a pea downwards. But little variation is seen in the eggs of this species. Average measurement, 2·8 inches in length, by 1·8 inch in breadth. Incubation, performed chiefly if not entirely by the female, lasts, it is said, about a month.

DIAGNOSTIC CHARACTERS: The elongated form and dark decided spots characterize the eggs of this and the following species (the only two known to breed in our area), and the small size is the distinguishing feature of those of the present bird. Unfortunately, however, the egg measurements of these two Divers overlap in rare instances, so that it behoves the collector thoroughly to identify his specimens.

Family COLYMBIDÆ. Genus COLYMBUS.

BLACK-THROATED DIVER.

COLYMBUS ARCTICUS, *Linnæus*.

Single Brooded. Laying season, latter half of May, and in June.

BRITISH BREEDING AREA: The Black-throated Diver is a much rarer and more local species than the preceding, nevertheless it is known to breed in various parts of the Outer Hebrides, and on the mainland from Argyll, through the counties of Inverness, Ross, and Sutherland, to Caithness. It has not yet been known to breed in Ireland, and is only a winter visitor to England.

BREEDING HABITS: This Diver very closely resembles the preceding species in its habits and economy. It is a resident in our islands, but subject to considerable local and southern movement during autumn and winter. Its breeding-haunts are wild and secluded moorland and mountain lochs, pools, and tarns, preference being shown for such that contain islands. The Black-throated Diver is not a gregarious species during the nesting season, but several pairs may frequently be found within a comparatively small area, each, however, keeping to themselves and to one particular haunt. It is probable that this Diver also pairs for life, and shows much attachment to certain nesting-places. The nest, invariably made upon the ground, is generally placed at no great distance from the water's edge on an island, more or less covered with grass and other coarse vegetation. It is, however, usually made on the bare shingly beach rather than amongst the herbage. This nest is rather more substantial than that of the Red-throated Diver, a mass of stalks, roots, and drifted aquatic vegetable fragments, lined with grass. Instances, however, are on record where the eggs have been found on the bare earth. One would think that this Diver must occasionally remove its eggs in the event of a sudden rise of the water, so closely are some of the nests made to the margin of the pool. The bird is not a close sitter, slipping quietly off the moment she is alarmed, and taking refuge in the water, where she is soon joined by her ever-watchful mate. Sometimes the birds will fly round and scan the scene from the air.

RANGE OF EGG COLOURATION AND MEASUREMENT: The eggs of the Black-throated Diver are two in number, narrow and elongated in shape. They vary from olive-brown to rufous or buffish-brown in ground colour, sparingly spotted and speckled with blackish-brown, and with a few obscure underlying markings of paler brown. The spots range from the size of a pea downwards, and are distributed here and there over the surface, and are generally more numerous towards the larger end of the egg. Average measurement, 3·2 inches in length, by 2·0 inches in breadth. Incubation, performed chiefly if not entirely by the female, is said to last a month.

DIAGNOSTIC CHARACTERS: The eggs of this Diver are best distinguished by their larger size and less profuse spotting, but inasmuch as they overlap (rarely) in measurement with those of the preceding species, careful identification is imperative for perfect accuracy.

Family PODICIPEDIDÆ. Genus PODICEPS.

LITTLE GREBE.

PODICEPS MINOR (*Brisson*).

Double Brooded. Laying season, end of March to early May and July.

BRITISH BREEDING AREA: The Little Grebe is commonly and widely distributed throughout the British Islands in all districts suited to its requirements, with the exception of the Shetlands, to which it is only a wanderer in winter.

BREEDING HABITS: The Little Grebe is a resident in the British Islands, but subject to considerable local and southern movement during winter. Its breeding-haunts are reed- and rush-fringed pools, canals, and slow-running streams; it is remarkable what a small sheet of water will in many instances content a pair of these interesting little birds. It may be found on almost every description of water, provided always that there is considerable cover in the shape of rushes, reeds, equisetums, and the like, growing in the shallows, and long coarse grass, tall plants, and weeds of all kinds flourishing on the banks. Although several pairs may occasionally be seen swimming at no great distance, even on the same sheet of water, the Little Grebe is not at all gregarious during the breeding season, and each pair keeps to a favourite spot, resenting any intrusion of their neighbours. The nest is usually a more or less floating structure, in most cases built up from the bottom, and is made amongst the reeds, often at some distance from the bank at the edge of a patch of rushes or iris. Less frequently it is made amongst the vegetation growing on the banks of the water, more often just in the water, and partly concealed by overhanging plants and grass. Nests have been recorded made in the low branches overhanging the water. It may best be described as a mass of half-rotten, wet vegetation, reeds, rushes, equisetums, grass, and water-weeds all matted together, with a shallow hollow at the top more neatly finished. The bird sits lightly, at the least alarm quickly covering her eggs with bits of wet weed and rush to conceal them from view, and then dropping so quietly into the water that scarcely a ripple remains to disclose the secret of her disappearance. It is amazing how quickly this operation is performed; and here once more I might impress upon the reader that this act of covering the eggs is *not* for warmth, but for concealment. If the bird is disturbed suddenly, and has to leave the eggs uncovered, it usually returns at the first opportunity and does so.

RANGE OF EGG COLOURATION AND MEASUREMENT: The eggs of the Little Grebe are from four to six, rarely seven, in number, elliptical and pointed in shape, and rather rough in texture. They are pure white when newly laid, but soon become stained and discoloured by contact with the wet nest, and feet and plumage of the parents. Average measurement, 1·5 inch in length, by 1·0 inch in breadth. Incubation, performed by both sexes, lasts about three weeks.

DIAGNOSTIC CHARACTERS: The small size, elliptic shape, and pea-green tinge of the interior of the shell viewed through the hole when held up to the light, readily distinguish the eggs of the Little Grebe from those of all other species breeding within our area.

Family PODICIPEDIDÆ.   Genus PODICEPS.

GREAT CRESTED GREBE.

PODICEPS CRISTATUS (*Linnæus*).

Single Brooded.   Laying season, April, May, and June.

BRITISH BREEDING AREA: The Great Crested Grebe is by far the most common in England and Wales, only breeding in one or two localities in Ireland, and even much more sparingly still in the south of Scotland. Its principal breeding-places in England are the broads of Norfolk and Suffolk, and on suitable waters in Yorkshire, Lancashire, Cheshire, Notts, Shropshire, Warwickshire, Oxfordshire, Bucks, and Hertfordshire. It breeds on the larger lakes of Wales, notably in Breconshire.

BREEDING HABITS: The Great Crested Grebe is a resident in our islands, but more widely dispersed in winter than in summer. Its breeding-haunts are the more extensive sheets of still water, the bird seldom or never frequenting such small pools as so often and generally content the Little Grebe. The bird is more or less gregarious during the breeding season, numbers of nests often being made within a small area, and their owners swimming in company all the summer through. The size of the colony of course entirely depends on the suitability of the waters and the abundance or rarity of birds in the vicinity. I am of opinion that this Grebe pairs for life, and yearly resorts to certain favoured places to breed. Early in spring, generally in March, the birds appear in their accustomed nesting-places, swimming in pairs, the male occasionally gambolling with and chasing the female, and both frequently engaging in various grotesque antics as preliminary to the season of more important courtship. The nest is almost invariably made in the water, either floating and moored to the reeds, or built up from the bottom of the pool in the same shelter, the matter being purely a question of the depth of water. It is a huge heap or raft of dead aquatic muddy vegetation of all kinds, with a shallow depression at the top more neatly finished, in which the eggs are laid. Several mock nests are generally to be found in its immediate neighbourhood, made whilst incubation is in progress, and most probably destined for the use of the young as resting-places. The bird sits lightly, covering its eggs (but not so effectually as the preceding species) with pieces of wet weed, and slipping off them into the water the moment it is alarmed. This Grebe, even though gregarious, resents intrusion, and will attack and drive off any other Grebes that may trespass on its own particular retreat.

RANGE OF EGG COLOURATION AND MEASUREMENT: The eggs of the Great Crested Grebe are from three to five in number, very rarely six, and most generally three. They are white when newly laid, more or less chalky in appearance, elliptical, and pointed at each end. When held up to the light the interior of the shell is delicate green. Contact with the wet nest and the feet and plumage of the bird soon robs them of their purity, so that specimens without ochre-tinted stains are somewhat difficult to get. Average measurement, 2·2 inches in length, by 1·5 inch in breadth. Incubation, performed by both sexes, may be about a month, but nothing definite appears to have been decided.

DIAGNOSTIC CHARACTERS: The eggs of the Great Crested Grebe may be distinguished from those of any other species breeding in our area by their size, colour, and elliptical shape, usually as much pointed at one end as the other, or nearly so.

There is strong evidence that the Black-necked Grebe, *Podiceps nigricollis*, Brehm, has bred in our islands, in Norfolk, but up to the present time no nest has been actually discovered. Booth records (*Rough Notes*) that he had an old bird with two downy nestlings brought to him by a marshman, but it is most extraordinary that such a painstaking and observant naturalist (thoroughly well aware of the importance of the discovery) should fail to remember either the date or the locality of the occurrence! There can be little doubt that this species would breed with us more or less regularly if not so persecuted in the spring, when its beautiful nuptial plumage attracts the fatal attention of the sportsman and the collector (or rather slayer) of rare birds. The nesting habits of this bird resemble those of allied species. The eggs are laid in May and June, and are four or five in number, creamy white when newly laid, but soon become stained. Average measurement, 1·8 inch in length, by 1·15 inch in breadth.

The Sclavonian Grebe, *Podiceps cornutus* (Gmelin), is said to nest sparingly near some fresh-water lochs at Gairloch in Ross-shire, but until better evidence is forthcoming I must decline to admit this species into the present work. Its nesting economy is similar to that of allied species. The eggs, laid in May and June, are four or five in number, and practically indistinguishable from those of *P. nigricollis*.

Family RALLIDÆ. Genus CREX.

CORN CRAKE.

CREX PRATENSIS, *Bechstein.*

Single Brooded. Laying season, end of May and in June.

BRITISH BREEDING AREA : The Corn Crake is widely and generally distributed throughout the British Islands, including the Hebrides, the Orkneys, and the Channel Islands.

BREEDING HABITS : The Corn Crake is a summer migrant to the British Islands, reaching them in April and May, but a few individuals winter in them, probably birds from more northern localities. The favourite breeding-haunts of this familiar bird are grass lands, especially hay-meadows, and fields of growing grain, pulse, and clover. I have also, in Devonshire, remarked its partiality for osier-beds, especially such as are clothed with a good undergrowth of rank grass and weeds. The Corn Crake is neither gregarious nor social during the breeding season, although I have known two nests in one field on several occasions; in one instance two were close together, only a few yards between them. This species pairs annually. The nest is invariably made on the ground, usually amongst clover or mowing grass, and far less frequently amongst grain or pulse. I have known it in osier-beds. It is a well-made structure, made externally of coarse and dry stems of herbage and a few dry leaves, and lined neatly with fine grass, much of it semi-green. The bird sits closely, often losing its life, as it broods over its eggs, from an unlucky stroke of the scythe or modern mowing-machine. I have known this species remove its eggs to a safer position after the grass has been cut, or when much disturbed by frequent visits to the nest.

RANGE OF EGG COLOURATION AND MEASUREMENT: The eggs of the Corn Crake are from eight to twelve in number. They vary from yellowish white to pale buff and pale blue in ground colour, spotted, speckled, and blotched with reddish-brown of various shades, and with underlying markings of pale violet-gray and purple. They are subject to considerable variation, and sometimes one egg of the pale blue ground colour type will be found amongst a clutch of the normal hue. The markings are not very profuse nor very large generally, and fairly well distributed over the entire surface of the shell. Average measurement, 1·4 inch in length, by 1·1 inch in breadth. Incubation, performed chiefly by the female, lasts three weeks.

DIAGNOSTIC CHARACTERS: The size of the egg and character of the markings readily distinguish the eggs of the Corn Crake from those of all other species breeding within our area, with perhaps the sole exception of those of the Water Rail. As a rule the eggs of the latter bird are more sparingly marked; whilst the breeding-grounds of the species are very different in character.

Family RALLIDÆ.   Genus CREX.

SPOTTED CRAKE.

CREX PORZANA, *Linnæus*.

Single Brooded.   Laying season, May and June.

BRITISH BREEDING AREA: The Spotted Crake is a somewhat scarce and local species, although widely distributed. It perhaps breeds most frequently in the low-lying eastern counties between the Humber and the Thames; less commonly it is however known to do so in Durham, Northumberland, Cumberland, and Notts. In Wales among other localities the marshes of Breconshire may be mentioned. North of the Border it breeds in Dumfries-shire, and thence locally through the eastern counties to at least as far north as the Moray Firth. In Ireland it has been known to breed in Co. Roscommon, and probably does so in Co. Kerry.

BREEDING HABITS: The Spotted Crake is a summer migrant to the British Islands, arriving in May, but a few individuals winter with us, probably from more northern lands. The breeding-haunts of this bird are marshes and the reed- and sedge-fringed margins of lakes, broads, and pools. It is a most skulking species, undoubtedly much overlooked, and regarded as rarer than it really is. In spite of the fact that numbers of nests may be found within small area, especially where the birds are plentiful, I do not think the Spotted Crake can be classed as gregarious or even social during the breeding season. It appears to pair annually, and to keep much to one particular haunt until the young are hatched. The nest is usually made amongst reeds, or in the centre of a hassock of sedge. It is a bulky structure standing well above the water, and built up from the bottom. It is made of decaying flags, reeds, bits of sedge, and other aquatic herbage, and the shallow cavity at the top is more neatly lined with dry grass and bits of finer reed and flag. The bird sits rather closely, but generally manages to slip off the nest unseen before the intruder reaches its immediate vicinity. It makes little or no demonstration at the nest.

RANGE OF EGG COLOURATION AND MEASUREMENT: The eggs of the Spotted Crake are from eight to twelve in number. They vary from white, or white tinged with green, to buff in ground colour, spotted and speckled with reddish-brown of various shades, and with underlying markings of violet-gray. The spots, which vary from the size of a pea downwards to mere dust-like specks, are pretty generally distributed over the entire surface, and the gray underlying marks on some eggs are large and predominate, or are small and few in number. Average measurement, 1·35 inch in length, by ·9 inch in breadth. Incubation, performed by both parents, lasts three weeks.

DIAGNOSTIC CHARACTERS: The eggs of the Spotted Crake are readily distinguished from those of all other species breeding within our area by the large size and boldness of the spots, and by the tinge of green which shows in the interior of the shell when held up to the light.

Baillon's Crake (*Crex bailloni*) has certainly bred in Norfolk and Cambridgeshire, its nest, containing eggs, having twice been obtained in both these counties. There can however be little doubt that the bird is now extinct as a regular breeding species, and the occurrence therefore is only of historical interest. The breeding season of this Crake begins in May, and as two broods appear to be reared in the season, fresh eggs may again be found up to August. The nest, built on the ground, is usually placed amongst reeds or sedge in the fens and swamps, and is made of dry leaves, sedges, and bits of broken reed, and lined with dry grasses. The eggs are from five to eight in number, and vary in ground colour from olive-brown to pale buff, mottled and sprinkled over the entire surface with indistinct and ill-defined markings of darker olive-brown, and with a few underlying markings of gray. Average measurement, 1·1 inch in length, by ·8 inch in breadth. The eggs of Baillon's

Crake very closely resemble those of the Little Crake, and require the most careful identification. Incubation, performed chiefly by the female, lasts from twenty-one to twenty-three days.

Family RALLIDÆ. Genus RALLUS.

WATER-RAIL.

RALLUS AQUATICUS, *Linnæus*.

Probably Double Brooded. Laying season, April, May, June, and July.

BRITISH BREEDING AREA: The Water-Rail is widely and generally distributed throughout the British Islands in all districts suited to its requirements, including the Hebrides, the Orkneys, and the Shetlands, but can nowhere be regarded as abundant with the solitary exception perhaps of the Norfolk Broads.

BREEDING HABITS: The Water-Rail is for the most part a resident in the British Islands, but subject to much local movement during winter, and there is a perceptible arrival of individuals from more northern haunts in autumn, especially in East Anglia. The favourite breeding-haunts of this very shy skulking Rail are pools and stagnant waters which contain an abundant growth of reeds, osier-beds with a dense thick undergrowth of coarse vegetation, especially when near to open water, and boggy ground overgrown with thickets of reeds. The Water-Rail is not gregarious, and each pair keeps closely to its own particular haunt. It is probable that this bird pairs for life. The nest is frequently made in a small thicket of reeds, or under the arching shelter of a clump of sedge or rushes, and in most cases is extremely difficult to find. It is rather bulky, and made of stems and flat leaves of reeds, and bits of dead aquatic plants, neatly and smoothly lined with finer yet similar material. It is invariably placed upon the ground, but when in a hassock of sedge or rushes the roots of the plant elevate it a foot or more above the surrounding level. The bird sits closely, and leaves the nest in a quiet, stealthy manner, slipping off into the surrounding vegetation and skulking close until the disturbance has passed.

RANGE OF EGG COLOURATION AND MEASUREMENT: The eggs of the Water-Rail are usually from five to seven in number, but as many as nine and more rarely eleven have been found. They vary from yellowish-white to pale buff in ground colour, somewhat sparingly spotted and speckled with reddish-brown and with numerous underlying markings of violet-gray. Little variation is presented, but occasionally the spots are rather large. Average measurement, 1·4 inch in length, by 1·0 inch in breadth. Incubation, performed by both sexes, lasts about three weeks.

DIAGNOSTIC CHARACTERS: The eggs of the Water-Rail cannot readily be confused with those of any other species breeding in our area, with the one exception of those of the Corn Crake; the markings, however, are smaller and more clearly defined; whilst the breeding-grounds of the two birds are totally different in character.

Family RALLIDÆ. Genus GALLINULA.

WATERHEN.

GALLINULA CHLOROPUS (*Linnæus*).

Double Brooded.   Laying season, March to July.

BRITISH BREEDING AREA: The Waterhen is commonly and widely distributed throughout the British Islands, even extending to such bare and wild localities as the Outer Hebrides and the Orkneys, but not reaching the Shetlands as a breeding species, only as a wanderer.

BREEDING HABITS: The Waterhen is a resident in the British Islands, and for the most part sedentary, unless driven out by long-continued frosts. It may be found breeding on the banks of almost every description of water, provided shelter of some kind is available; whilst in many places it lives almost in a state of semi-domestication. We can scarcely class the Waterhen as gregarious, but it is certainly to a great extent social during the breeding season, numbers of nests often being made within a small area, yet even then each pair shows a strong disposition to resent encroachment on its own particular nest-haunt, though ready enough to swim and feed in company with the rest. I am of opinion that this bird pairs for life, and not only keeps to one haunt season by season, but often makes its nest in one particular spot. The nest is placed in a great variety of situations, perhaps most frequently among flags, rushes, reeds, and iris, often at some distance from shore, in moderately shallow water. Sometimes it is built amongst a mass of branches bent down into the water, and is then entirely supported by the network of twigs; at others it is made amongst exposed roots, on the banks of the water. More rarely a fir tree has been selected, the nest being made as much as twenty feet from the ground. It is a bulky structure of rotten aquatic vegetation, flags, reeds, rushes, weeds, sedges loosely interwoven, but well massed together, the cavity containing the eggs being shallow, and lined with finer and drier material. The nest is often increased during the progress of incubation, either to repair the damages caused by the incessant wash of the waves, or to prepare for sudden floods. The bird does not sit very closely, and covers her eggs before leaving them, slipping off into the water at the first alarm.

RANGE OF EGG COLOURATION AND MEASUREMENT: The eggs of the Waterhen are from six to ten in number, sometimes as many as twelve. They vary from buffish-white to pale reddish-buff in ground colour, spotted and speckled with reddish-brown, and with underlying markings of gray. On some varieties the markings are of the character of blotches rather than spots, occasionally of considerable size; on others the markings are few and small. They are pretty evenly distributed over the surface. Average measurement, 1·7 inch in length, by 1·2 inch in breadth. Incubation, performed by both sexes, lasts from twenty to twenty-four days.

DIAGNOSTIC CHARACTERS: The size and general colouration of the eggs of the Waterhen prevent them from being confused with those of any other species breeding within our area.

Family RALLIDÆ.　　　　　　　　　　　　　Genus FULICA.

COMMON COOT.

FULICA ATRA, *Linnæus.*

Double Brooded.　　Laying season, May and July.

BRITISH BREEDING AREA: The Common Coot, although nowhere perhaps as abundant as the Water-hen, is nevertheless very generally dispersed over the British Islands, breeding in every county in all suitable localities, including the Hebrides and the Orkneys, but is only a wanderer to the Shetlands.

BREEDING HABITS: ¡The Coot is a resident in the British Islands, but subject to much local movement during autumn and winter, when its numbers are increased by individuals from more northerly areas. The breeding-haunts of the Coot are large ponds, reservoirs, broads, lakes, and slow-running rivers and streams; but the bird is not seen on such small expanses of water as so often content the preceding species. It shows a preference for broad open sheets of water, especially those that contain a good growth of reeds, equisetums, and such-like plants. The Coot can scarcely be regarded as gregarious during the breeding season, but in many localities great numbers of birds nest in a small area, each pair, however, keeping to a chosen spot, and resenting intrusion. It is social notwithstanding, and often swims and feeds in company during this period. It is probable that this species pairs for life; certain places are tenanted every year, and the birds seem much attached to their haunts. The nests are placed in a variety of situations. Sometimes they are made amongst reeds, rushes, or iris, in shallow water, at a considerable distance from the shore; sometimes a small island clothed with grass and other coarse herbage is selected; frequently the nest is a floating structure, moored securely to reeds or flags; less often it may be found amongst the tangled vegetation on the dry bank, but close to the water's edge. It is usually a bulky structure, the foundation often built at the bottom of the water, and then carried upwards from a few inches to a foot above the surface. Most of the nest is a mere heap of rotten aquatic vegetation of all kinds, at the top of which a shallow cavity is formed, and lined with finer and drier material. The Coot sits lightly, and when alarmed slips very quietly off the nest into the water, and skulks amongst the vegetation until all is safe again; but the bird never attempts to conceal the eggs by covering them.

RANGE OF EGG COLOURATION AND MEASUREMENT: The eggs of the Coot are from six to twelve in number, seven or eight being an average clutch. They are buffish-white or very pale clay-buff in ground colour, sprinkled, speckled, and dusted over most of the surface with dark blackish-brown, and with underlying markings similar in character of violet-gray. But little variation is presented, except in the tint of the ground colour. Average measurement, 2·1 inches in length, by 1·3 inch in breadth. Incubation, performed by both sexes, lasts from twenty-one to twenty-three days.

DIAGNOSTIC CHARACTERS: The eggs of the Coot are readily distinguished from those of all other species breeding in our area, by their size and fine markings sprinkled over most of the surface.

Family COLUMBIDÆ.   Genus COLUMBA.

RING-DOVE.

COLUMBA PALUMBUS, *Linnæus*.

Double Brooded.   Laying season, March to September;
even later.

BRITISH BREEDING AREA: The Ring-Dove is a common and very widely distributed species, breeding throughout the wooded or fairly well timbered and cultivated districts of the British Islands. This bird has increased its area considerably within a comparatively recent period, following the planting and growth of trees. It would also appear to have increased both by chronic and irruptic emigration.

BREEDING HABITS: The Ring-Dove is a resident in the British Islands, but largely increased in numbers during autumn and winter by birds from the Continent. Its favourite breeding-grounds are woods, parks, plantations, shrubberies, and well-timbered lands. It not only breeds in the London parks in gradually increasing numbers, but also frequents wooded pleasure-grounds in many of our smaller towns. Numbers do so in the various wooded grounds of Torquay, for instance. During the non-breeding season the Ring-Dove is certainly gregarious to a very great extent, and even during the nesting period, which lasts practically all the spring and summer, a varying amount of sociability may be remarked. The nests, however, are never made in colonies. I am of opinion that this bird pairs for life. The nest is placed in a great variety of situations, both in evergreen and in deciduous trees (often in the latter before they are in leaf), in large bushes, or amongst ivy on cliffs and tree-trunks.   Woods, plantations—

especially those of fir and pine—isolated trees in the hedgerows or in the open fields, or even tall dense hedges, are all selected without any very appreciable choice or selection. The nest is made at almost any height, from a few feet to a hundred from the ground. It is a simple structure, merely a few dead sticks and twigs wove basket-like into a flat platform, through which the eggs may often be seen from below. I have known the nest of this bird to be made in the abnormal bush-like growths so common on the wild cherry tree and the elm, also in tufts of mistletoe on poplars and whitethorns. The bird is not a very close sitter, usually dashing from the nest at the first disturbance, although I have known it remain on the eggs very quietly and persistently if it thought itself unseen. It makes little or no demonstration, and flies right away into the cover.

RANGE OF EGG COLOURATION AND MEASUREMENT: The eggs of the Ring-Dove are two in number normally, but very exceptionally three, or even one. They are oval and elongated in form, smooth in texture, and with some polish, and pure white. Average measurement, 1·6 inch in length, by 1·25 inch in breadth. Incubation, performed by both sexes, lasts from seventeen to twenty days.

DIAGNOSTIC CHARACTERS: The eggs of this species can only be confused with those of the Rock-Dove (excepting of course those of the Domestic Pigeon), but they are almost invariably larger. The two birds also breed in widely different situations (conf. p. 349).

Family COLUMBIDÆ. Genus COLUMBA.

STOCK-DOVE.

COLUMBA ŒNAS, *Brisson.*

Double Brooded. Laying season, April to August and September; even later.

BRITISH BREEDING AREA: The Stock-Dove is not perhaps so abundant as the Ring-Dove, and is more locally distributed, nevertheless its numbers not only appear to be steadily on the increase, but its area is gradually expanding, especially northwards. It breeds pretty generally throughout England and Wales, both inland and near the coast; and is now known to do so, if rarely, in Scotland, notably in Stirlingshire and along the shores of the Moray and Dornoch Firths. The bird also appears slowly to be establishing itself in Ireland, where at present it is known to inhabit the extreme north-eastern portions of the island.

BREEDING HABITS: The Stock-Dove is a resident in the British Islands, and is almost everywhere, especially in littoral districts, confused with the Rock-Dove, from which, it need scarcely be pointed out, it may instantly be distinguished, even in the air, by having the rump uniform in colour with the back, and the wing bars broken and rudimentary. Its haunts are the wooded districts, especially the old forest areas where most of the timber is aged and full of hollows and crevices. The bird may also be found in quarries, even in treeless districts, and on wooded sea-crags, and even amongst ocean cliffs. This is especially the case in Devonshire, the Isle of Wight, the cliffs of Dorset, and Flamborough Head in Yorkshire. In other localities partiality is shown for warrens and links. This bird is gregarious enough during autumn and winter, and even during the nesting season is remarkably social, in some places breeding in colonies, as for instance in Sherwood Forest. I am of opinion that the Stock-Dove pairs for life, yearly returning to certain haunts to breed, and being much attached to its nesting-places. The nest is always placed in a covered site by preference; and, so far as my experience of this species extends, holes in trees and cliffs, and the deserted nests of Crows and Magpies, and the old dreys of squirrels are the favourite situations. It is, however, frequently made amongst dense masses of ivy, in pollard trees, and in rabbit-earths, and occasionally in church steeples and old towers. It is a slight structure—a mere mat of twigs, roots, or straws, carelessly arranged, and in many cases is dispensed with altogether. The bird is rather a close sitter; but this depends a good deal on the situation of the nest. It makes no demonstration when disturbed, and flies right away to some safe retreat.

RANGE OF EGG COLOURATION AND MEASUREMENT: The eggs of the Stock-Dove are normally two in number, but three have been recorded. They are creamy-white in colour, oval and elongated in form, and smooth in texture. Average measurement, 1·4 inch in length, by 1·2 inch in breadth. Incubation, performed by both sexes, lasts from seventeen to eighteen days.

DIAGNOSTIC CHARACTERS: The size and yellowish tinge of the eggs of the Stock-Dove readily distinguish them from those of all other allied species breeding in our islands.

Family COLUMBIDÆ.   Genus COLUMBA.

ROCK-DOVE.

COLUMBA LIVIA, *Brisson.*

Double Brooded.   Laying season, March and April to September.

BRITISH BREEDING AREA: The Rock-Dove is more or less commonly distributed along such rocky coasts as are suited to its requirements throughout the British Islands. It is sparingly and locally distributed on the eastern and southern coasts of England, but becomes abundant further north, especially in the Hebrides, the Orkneys, and the Shetlands, and along the western shores of the Scottish mainland; whilst the same remarks apply to Ireland, the bird being specially numerous on all the wave-hollowed cliffs. The inland colonies of this bird are unquestionably Domestic Pigeons that have become feral, or their descendants.

BREEDING HABITS: The Rock-Dove is a resident in the British Islands. The truly wild Rock-Dove is only found on the coast. Its favourite breeding-haunts are such parts of the marine precipices that abound with fissures and wave-worked tunnels and caves. All through the year it is more or less gregarious, and may generally be found breeding in colonies of varying size, according to the amount of accommodation afforded. The Rock-Dove prefers a cave, or a large and roomy hollow in the cliffs, which is either washed by the sea, or only accessible to the most daring of climbers. If caves are not to be had, the colonies are more scattered up and down the cliffs, wherever suitable fissures and clefts can be found. This bird pairs for life, and continues to resort to certain caves year after year; some of its retreats being famed from remote times for the numbers of Doves frequenting them. The nest is placed on ledges, or in cracks and crannies in the roof, or on any little prominence on the rugged sides of the cave, whilst those on the cliffs are deep in the fissures. It is a simple structure, rudely put together, and made of dry grass, sea-weed, twigs, roots, or the dry stems of weeds; whilst grass in a green state has been known to be used. The birds sit pretty closely, dashing out of the caves and fissures in twos and threes as their haunt is approached and explored, and flying right away with no demonstration whatever.

RANGE OF EGG COLOURATION AND MEASUREMENT: The eggs of the Rock-Dove are two in number, somewhat short and oval in form, and pure white. Average measurement, 1·4 inch in length, by 1·2 inch in breadth. Incubation, performed by both sexes, lasts from sixteen to eighteen days.

DIAGNOSTIC CHARACTERS: The eggs of the Rock-Dove are not easily confused with those of any other species when taken direct from the nest, but otherwise they cannot with certainty be distinguished from those of the Ring-Dove, although they are usually smaller and rounder.

Family COLUMBIDÆ. Genus TURTUR.

TURTLE-DOVE.

TURTUR AURITUS, *Gray*.

Frequently Double Brooded. Laying Season, May and June and August.

BRITISH BREEDING AREA: North of the Humber and the valley of the Don, the Turtle-Dove is a somewhat rare and local bird, and is not known with certainty to breed in Scotland. South of the above-mentioned limits it is generally distributed throughout England and Wales, but becomes rarer again in the latter, and also in the extreme south-western districts of England. In Ireland it is only sparingly distributed in the well-wooded areas.

BREEDING HABITS: The Turtle-Dove is a summer migrant to the British Islands, arriving early in May, occasionally at the end of April. Its favourite breeding-haunts are woods, plantations, parks, pleasure-grounds, and well-timbered agricultural districts, especially those in which the hedges are tall and dense. It is not gregarious during the nesting period, neither is any marked social tendency observable. I am of opinion that this species pairs for life. I saw it in pairs even whilst on migration in Algeria, and certain haunts in our islands are visited yearly with marked regularity. The nest is seldom at such a great altitude as that of the Ring-Dove so often is, being more usually built in tall bushes rather than trees. A thick dense hedge is a favourite situation, and the nest may commonly be found in small whitethorns, tall hollies, and even laurels. It is a flat, mat-like structure, made with a few slender dead twigs woven like basket-work, and is so slight that the eggs may often be seen through it from below. The bird is a rather close sitter, makes no demonstration when disturbed, and usually hides itself as soon as possible amongst the nearest trees or bushes.

RANGE OF EGG COLOURATION AND MEASUREMENT: The eggs of the Turtle-Dove are two in number, oval in form, and almost as much pointed at one end as the other, smooth in texture, and creamy-white in colour. Average measurement, 1·2 inch in length, by ·91 inch in breadth. Incubation, performed by both sexes, lasts about sixteen days.

DIAGNOSTIC CHARACTERS: The eggs of the Turtle-Dove may be readily distinguished from those of all other species breeding in our area by their size, form, and creamy-white tint.

Not the least interesting phase of the last great irruption of Pallas's Sand Grouse (*Syrrhaptes paradoxus*) into the British Islands during 1888, was the fact of some of the individuals attempting to breed. There is strong evidence that the eggs of this interesting bird were taken or seen in Yorkshire and Norfolk, and on the coast of the Moray Firth, whilst the nestlings were both observed and captured in the latter district. There can be little doubt however that the Sand Grouse will never succeed in establishing itself in our islands, and its nesting in them is thoroughly abnormal. The nest of this Sand Grouse is merely a hollow in the sandy ground, lined with a few bents or dead leaves. The eggs are usually three, but exceptionally four, in number, very oval and Pigeon-egg like in shape, and olive or brownish-buff in ground colour, profusely spotted with dark brown, and with underlying markings of violet-gray. Average measurement, 1·7 inch in length, by 1·1 inch in breadth. Incubation, performed by the female, lasts about a month. It is said that this species rears two broods in the year, the eggs for the first being laid in April. The eggs of this Sand Grouse cannot readily be confused with those of any other species breeding in our area, their oval form being very characteristic.

Family PHASIANIDÆ. Genus COTURNIX.
Sub-family *PERDICINÆ*.

COMMON QUAIL.

COTURNIX COMMUNIS, *Bonnaterre*.

Single Brooded.   Laying season, end of May and in June.

BRITISH BREEDING AREA: The Quail can nowhere be regarded as an abundant bird in our islands, but is generally distributed throughout their area, even including such wild districts as the Outer Hebrides, the Orkneys, and the Shetlands, becoming most numerous in the southern and central counties of England.

BREEDING HABITS: It is most probable that the individual Quails that breed in our islands are summer migrants, the birds that winter with more or less regularity in the south and west of England and in Ireland being wanderers from more northern lands. The Quail arrives in our islands in May. Its favourite breeding-grounds are grain lands, hay and clover meadows, and unenclosed areas of rough pasturage. It is much attached to certain haunts, returning to them every season to breed; and it is probably owing to this attachment and to the drainage of so much rough pasture that it has so perceptibly decreased in numbers during recent years. In some districts the Quail is polygamous, in others monogamous, the excess or rarity of hens apparently

PLATE XII

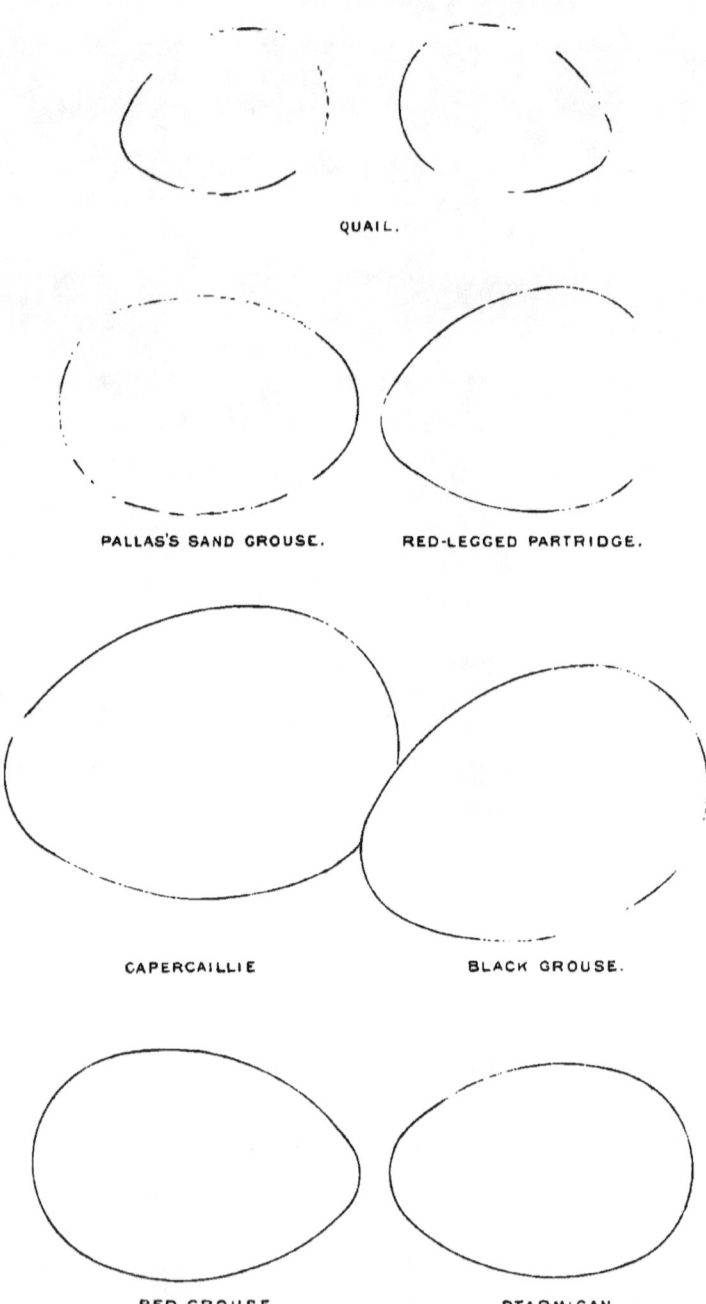

determining the matter. During the pairing season the cocks are very pugnacious, even in localities where the birds are few, and constant flights are taking place until the hens have gone to nest. In places where polygamous instincts prevail, and where a cock runs with several hens, the nests may be found not many yards apart, two females occasionally sharing the same abode. The nest is a mere hollow, amongst growing grain or grass and clover, scantily lined with a little dry grass and a few dead leaves. The hen is a rather close sitter, especially where the herbage is tall and dense, and when alarmed slips very quietly and with no demonstration from the eggs into the surrounding cover.

RANGE OF EGG COLOURATION AND MEASUREMENT: The eggs of the Quail are from eight to twelve in number; I have known nests contain as many as twenty, but these were doubtless the produce of two hens. They are buffish-white or clear yellowish-olive in ground colour, boldly blotched and spotted with various shades of umber-brown and blackish-brown. There are two distinct types: one in which the markings are small and dark (spots not blotches), and sprinkled over the entire surface; the other in which the markings are bold, large, and irregular (blotches not spots) and often confluent. As a rule the pale ground colour is associated with the former type, the olive ground colour with the latter. Average measurement, 1·1 inch in length, by ·91 inch in breadth. Incubation, performed by the female, lasts about three weeks.

DIAGNOSTIC CHARACTERS: The size and the abundance of the markings, and the absence of any underlying spots readily distinguish the eggs of the Quail from those of any other species breeding in our area.

Family PHASIANIDÆ.   Genus CACCABIS.
Sub-family PERDICINÆ.

RED-LEGGED PARTRIDGE.

CACCABIS RUFA (*Linnæus*).

Single Brooded. Laying season, end of April and in May.

BRITISH BREEDING AREA: The Red-legged Partridge was introduced into England more than a century ago, but owing to its partiality for dry sandy soils, it still remains very local. It is fairly common wherever it is preserved, in the counties of Norfolk, Suffolk, Essex, Kent, and Sussex; and evidence is not wanting that it is endeavouring to establish itself in Lincolnshire, the Midlands, and the higher grounds on the north side of the Thames valley. The attempts to introduce this species into Scotland and Ireland have hitherto proved futile.

BREEDING HABITS: The Red-legged Partridge is a resident in those areas it affects. Its breeding-haunts not only embrace the cultivated fields, but rougher ground, such as commons, the open treeless parts of woodlands, low sedgy meadows, and strips of heathy land clothed with clumps of gorse and rushes and thickets of brambles and briars. The bird is not gregarious during the breeding season. The Red-legged Partridge pairs annually, and during the mating season the males become quarrelsome and pugnacious. The nest is made amongst the thick herbage of a hedge-bottom or a dry ditch, amongst growing grain, grass or clover crops, not unfrequently in an exposed situation by a public footpath, and occasionally amongst the thatch of a stack, or even in the side. It is a mere hollow scraped out by the female, and carelessly lined with a few bits of dry herbage and leaves. The bird sits closely, and when driven from the nest makes little or no demonstration, and generally flies right away at once.

RANGE OF EGG COLOURATION AND MEASUREMENT: The eggs of the Red-legged Partridge are from twelve to eighteen in number, and often laid at intervals of a few days. They vary from buff to brownish-yellow in ground colour, spotted and speckled with reddish-brown and chocolate-brown; the shell is strong and somewhat coarsely grained, but with some polish. The eggs of this bird are not subject to any great amount of variation, but some are more profusely spotted than others, and on some a few of the larger markings are irregular in shape and more blotchy. Average measurement, 1·6 inch in length, by 1·2 inch in breadth. Incubation, performed by the female, lasts twenty-four days.

DIAGNOSTIC CHARACTERS: The size of the egg and the character of the markings readily distinguish it from that of any other species breeding in our area.

Family PHASIANIDÆ.   Genus PERDIX.
   Sub-family *PERDICINÆ*.

COMMON PARTRIDGE.

PERDIX CINEREA, *Brisson*.

Single Brooded. Laying season, end of April to beginning of June.

BRITISH BREEDING AREA: The Common Partridge is generally distributed throughout the agricultural districts of the British Islands, and in a few more upland localities (as, for instance, near Dartmoor and other moorlands), wherever it is preserved or has been introduced.

BREEDING HABITS: The partridge is a resident in the British Islands. Districts most favourable to the preservation and increase of the Partridge are the well-cultivated farm-lands, where the fields are not too large, where the hedges are low and dense, and afford plenty of bottom cover, and where grain is grown in abundance. It also frequents many other localities, but does not thrive so well, such as on the borders of moorlands, on commons, and rough unenclosed ground. Although the Partridge is gregarious in autumn and winter, it becomes solitary during the breeding season, each pair keeping to themselves until the young are reared. It is monogamous and very probably pairs for life, but as many birds are widowed during the shooting season, a considerable amount of mating must take place annually. The nest is frequently made in a dry hedge-bottom or a ditch, amongst growing crops of all kinds, or in coarse vegetation on rougher ground. Nests are frequently made near public footpaths, close to a gate-post, or even on the top of a stack. The nest is a mere hollow, scantily lined with a little dry grass or other herbage, sometimes a few dead leaves. The bird sits closely, often allowing itself to be taken rather than leave its eggs, but when flushed makes little or no demonstration, and hurries into cover.

RANGE OF EGG COLOURATION AND MEASUREMENT: The eggs of the Partridge vary, according to the age of the hen, from ten to fifteen or twenty in number, but occasionally even larger clutches are found, as many as thirty-three having been recorded in a single nest. They are somewhat pyriform in shape, smooth in texture, with some polish, and generally pale olive-brown in colour: white and pale green varieties are, however, sometimes met with. Average measurement, 1·4 inch in length, by 1·15 inch in breadth. Incubation, performed by the female, lasts from twenty-one to twenty-four days.

DIAGNOSTIC CHARACTERS: The eggs of the Partridge may be readily distinguished, by their uniform olive tint and their size, from those of all other species breeding in our area.

Family PHASIANIDÆ.   Genus PHASIANUS.
Sub-family PHASIANINÆ.

PHEASANT.

PHASIANUS COLCHICUS, *Linnæus*.

Single Brooded. Laying season, April and May.

BRITISH BREEDING AREA: The Pheasant is widely and generally distributed throughout the British Islands, in all suitable districts where it is preserved, even extending to the Outer Hebrides, where it has been successfully introduced.

BREEDING HABITS: The habits of such a semi-domesticated species as the Pheasant need little description. The bird is of course a resident in our islands, where it would doubtless soon become extinct if the strict protection it now enjoys were withdrawn. Wherever it is fostered thus artificially by man it may be found, the only condition being the presence of cover. It thrives best in such woods as contain plenty of bottom growth, and adjoining which fields and open ground afford feeding-places. In a truly wild state the Pheasant appears to be strictly monogamous, but semi-domestication has so far affected its morals as to render it polygamous in our islands, one male running with several females. Towards the end of March the cocks begin to crow and fight for the possession of the hens, and about a month later the latter go to nest. On estates where the birds are common, many nests may be found within a small area, and all through the breeding season more or less social instincts prevail. The nest is almost invariably made on the ground, but instances are on record of its being found in stacks, and even in a disused squirrel's drey. It is often made amongst the undergrowth of the woods and plantations, or at the bottom of a hedge or dry ditch, or amongst growing grain and other crops. Sometimes it is made in the centre of a tuft of rushes, under brambles or heaps of cut brushwood. It is merely a hollow, lined with a few dead leaves, bits of withered bracken, or dry grass. The female sits closely, but when disturbed hurries from her charge into the nearest cover with little or no demonstration, but often with startling suddenness.

RANGE OF EGG COLOURATION AND MEASUREMENT: The eggs of the Pheasant are usually from eight to twelve in number, sometimes as many as twenty, and I have known twenty-six! They are smooth in texture, somewhat polished, but the shell is finely pitted, and vary from brown, through olive-brown, to bluish-green in colour. Occasionally one of the bluish-green type occurs among a clutch of the more usual tint. Average measurement, 1·8 inch in length, by 1·4 inch in breadth. Incubation, performed by the female, lasts about twenty-four days.

DIAGNOSTIC CHARACTERS: The eggs of the Pheasant may be readily distinguished by their uniform olive or brown tint, and their size, from those of all other species breeding in our area.

Family PHASIANIDÆ.   Genus TETRAO.
Sub-family TETRAONINÆ.

CAPERCAILLIE.

TETRAO UROGALLUS, *Linnæus*.

Single Brooded.   Laying season, April and May.

BRITISH BREEDING AREA: Considerable success has attended the re-introduction of the Capercaillie into Scotland, and doubtless the bird will ultimately extend its area into many districts both north and south of its present centre of distribution. Its head-quarters in our islands are the counties of Perth, Forfar, and Stirling.

BREEDING HABITS: The Capercaillie, as may scarcely be remarked, is a resident in the British Islands, but there is a strong tendency to chronic emigration, especially among young males. Its haunts are principally confined to the extensive fir, spruce, and larch forests, but it also frequents in smaller numbers birch and oak woods, and the rough, sparsely-timbered ground between the forests and the moorlands. The Capercaillie is polygamous, and during the mating season the males indulge in various antics, repairing in the morning and evening to certain stations or "leking-places," where after their display and love-cries are over, and the possession of the females decided by right of conquest, the males pair. Each cock runs with several hens, but takes no share in nesting duties. The nest is merely a hollow lined with a few dry leaves or a little withered grass and some pine-needles, made amongst the heather and bilberry wires in an open part of the forests, often beneath the shelter of a stunted bush, or sheltered by masses of fern and bramble. The bird is a close sitter, and when flushed flies off into cover with little or no demonstration.

RANGE OF EGG COLOURATION AND MEASUREMENT: The eggs of the Capercaillie are from eight to twelve in number, occasionally fourteen or even sixteen. They vary from brownish-buff to reddish-buff in ground colour, blotched sparingly, and spotted and speckled thickly with rich reddish-brown. The markings are generally small, rarely as big as a pea, and are usually distributed pretty evenly over the entire surface of the egg. Average measurement, 2·2 inches in length, by 1·6 inch in breadth. Incubation, performed by the female, lasts twenty-six days.

DIAGNOSTIC CHARACTERS: The size of the eggs, yellowish ground colour, and small distinct spots, distinguish them from those of all other allied species. They most closely resemble those of the Black Grouse, but may be instantly separated by their larger size.

Family PHASIANIDÆ.
Sub-family *TETRAONINÆ*.
Genus TETRAO.

BLACK GROUSE.

TETRAO TETRIX, *Linnæus*.

Single Brooded.   Laying season, April and May.

BRITISH BREEDING AREA: The Black Grouse is not so widely and generally distributed in England as was formerly the case, nevertheless it breeds locally in all the counties south of the Thames and the Bristol Avon, with the one exception of Kent. It is also locally distributed in Wales, the Midlands, and Norfolk, and also in every county north of Notts up to the Border. It is much more generally dispersed in Scotland, including some of the Inner Hebrides, but does not extend beyond the mainland northwards. It is not an inhabitant of Ireland.

BREEDING HABITS: The Black Grouse, of course, is a resident in the British Islands. Its favourite haunts are wild broken country near the moorlands, birch and fir plantations, and the rough valleys below the level plateaux of ling and heath, where the ground is clothed with bracken, gorse, and brambles, strewn with rock boulders, and traversed by trout streams that sometimes widen out into swamps covered with cotton-grass, rushes, and other coarse vegetation. In more lowland districts timbered commons and small tracts of moorland, surrounded by pine woods and plantations, with a good bottom growth, are its most attractive retreats. The Black Grouse is polygamous, and the males perform their courtship in a very similar manner to the preceding species, having regular "leking-places" to which they resort to show off and pair with as many females as their prowess or charms can ensure. The males take no further share in the domestic arrangements, but it is said they sometimes join the females and their broods. This must be very exceptional, however. The Gray Hen, as the female is generally called, makes a slight nest on the ground, under the shelter of a mass of withered bracken, or a heap of brambles, or amongst heath and ling, rushes, or bilberry wires. Sometimes it is made by the side of a boulder, or under a fallen log amongst long grass and fern. It is a mere hollow, carelessly lined with bits of fern, pine needles, dry grass, or dead leaves. The female is a close sitter, and when flushed hurries off into the nearest cover with little or no demonstration; but when the young are hatched her actions are very different and most alluring.

RANGE OF EGG COLOURATION AND MEASUREMENT: The eggs of the Black Grouse are usually from six to ten in number, but occasionally as many as sixteen may be found, probably the produce of two females. They vary from yellowish-white to brownish-buff in ground colour, spotted and blotched with reddish-brown of various shades. The shell is rather rough and granulated, but with some amount of polish. The spots are always much more numerous than the blotches, which are few (about the size of a pea) and often absent. Average measurement, 2·0 inches in length, by 1·4 inch in breadth. Incubation, performed by the female, lasts about twenty-six days.

DIAGNOSTIC CHARACTERS: The size of the eggs of the Black Grouse, their yellowish ground, and small brown markings, readily distinguish them from those of all other species breeding in our area.

Family PHASIANIDÆ. Genus LAGOPUS.
Sub-family *TETRAONINÆ*.

RED GROUSE.

LAGOPUS SCOTICUS (*Brisson*).

Single Brooded. Laying season, March to June, according to locality and weather.

BRITISH BREEDING AREA: East of an imaginary line drawn from the mouth of the Severn to the upper waters of the Humber, the Red Grouse is unknown. Its strongholds commence on the heaths of Wales, in Glamorganshire, and continue northwards through the moorlands of the Cambrian and Pennine mountain systems to the Border. North of the latter it is widely and generally distributed throughout Scotland as far north as the Orkneys, and as far west as the Outer Hebrides. Although not so common, it is equally widely distributed over the moorlands of Ireland. It is almost unnecessary to remark that this species is peculiar to our islands, being the island representative of the Willow Grouse.

BREEDING HABITS: The Red Grouse is, of course, a resident in the British Islands, and very closely confined to a certain area. Its exclusive haunts are the vast wastes of heath and ling that stretch almost continuously from South Wales to the Orkneys. This ground is broken and uneven enough, but almost destitute of timber; hills and dales, vast level plateaux, rolling plains and hollows, ridges and peaks, everywhere more or less luxuriantly covered with heath, with occasional expanses of swamp full of coarse grass, rushes, sedges, and the like, or broken areas in which gorse, bracken, broom, and various mountain fruits flourish. The Red Grouse pairs annually, early in spring, and though the males are pugnacious and demonstrative during this period, there is no evidence to show that any polygamous propensities are indulged in. We cannot class this Grouse as gregarious during the nesting season, but numbers of birds breed in close proximity, and a certain social tendency is frequently apparent during the spring and summer. The nest is always made upon the ground, usually among the ling and heather, less frequently in rushes or coarse grass, and generally under the shelter of a bush. Sometimes it is made by the side of a public footpath or highway, in a much-frequented spot. It is a mere hollow, carelessly lined with a little dry grass, or bits of withered heath and ling, or dead leaves. The bird is a close sitter, but when flushed makes little or no demonstration if the nest contains eggs only.

RANGE OF EGG COLOURATION AND MEASUREMENT: The eggs of the Red Grouse are from five to fifteen in number; in unfavourable seasons the clutches are small, in dry and warm ones large. They are creamy-white in ground colour, profusely and handsomely blotched and spotted with rich reddish or crimson-brown, or even blackish-brown. The markings are very numerous, and cover most of the shell; but as the colour, especially on newly-laid eggs, is very easily rubbed, much of the beauty is frequently destroyed by contact with the parent. Average measurement, 1·8 inch in length, by 1·25 inch in breadth. Incubation, performed by the female, lasts twenty-four days.

DIAGNOSTIC CHARACTERS: The size, rich colour, and profusion of the markings, readily distinguish the eggs of the Red Grouse from those of allied species, except perhaps from those of the Ptarmigan: the latter eggs, however, are more buff, and not so profusely marked.

Family PHASIANIDÆ.   Genus LAGOPUS.
Sub-family *TETRAONINÆ*.

PTARMIGAN.

LAGOPUS MUTUS (*Montin*).

Single Brooded. Laying season, May; late seasons, early June.

BRITISH BREEDING AREA: The Ptarmigan is another very local species, confined principally to the loftiest mountains of the Highlands, as far south as the mountains of Islay, Jura, and Arran, and extending westwards to Skye, Harris, and Lewis, but not to the Orkneys nor the Shetlands.

BREEDING HABITS: The Ptarmigan is resident in the British Islands, and subject to remarkably little local movement, except when driven to lower levels by prolonged bad weather. Its breeding-haunts are the bare wind-swept summits of the mountains, where the ground is broken and stony, rough with boulders, and only scantily clothed with ling, mosses, lichens, and various dwarf mountain fruits. The Ptarmigan is not at all gregarious during the breeding season, neither does it exhibit much social tendency during that period, the birds pairing early in spring, and each couple keeping much to themselves until the duties of the year are over. The nest, invariably on the ground, is usually made on the stone- and rock-strewn earth, sometimes under the lee of a large boulder, or partly hidden by a stunted plant. It is merely a hollow lined with a few twigs of heather, a little dry grass, or some dead leaves of the Alpine plants. The bird sits closely, sometimes until almost trodden upon, and is most difficult to see, so closely does her brown and gray plumage harmonize with surrounding tints. The eggs themselves are coloured in a very protective manner.

RANGE OF EGG COLOURATION AND MEASUREMENT: The eggs of the Ptarmigan are from eight to twelve in number. They vary in ground colour from creamy or grayish-white to brownish-buff, and are boldly blotched and spotted with rich liver-brown, sometimes almost black in intensity. There is but little variation amongst them to describe, but the markings are pretty evenly dispersed over the entire surface. Average measurement, 1·7 inch in length, by 1·1 inch in breadth. Incubation, performed by the female, lasts about twenty-one days.

DIAGNOSTIC CHARACTERS: The eggs of the Ptarmigan are readily distinguished by their size, buff ground colour, and large bold markings. They are somewhat similar to those of the Red Grouse, but never so profusely marked.

INDEX.

Accentor modularis, 146
Accipiter nisus, 212
Acredula caudata rosea, 83
Acrocephalus arundinaceus, 117
Acrocephalus palustris, 115
Acrocephalus phragmitis, 120
Ægialitis hiaticula major, 260
Ægialophilus cantianus, 259
Alauda arborea, 60
Alauda arvensis, 58
Alca torda, 313
Alcedo ispida, 181
Aluco flammeus, 182
Anas acuta, 229
Anas boschas, 237
Anas circia, 234
Anas clypeata, 235
Anas crecca, 232
Anas penelope, 230
Anas strepera, 227
Anser cinereus, 224
Anthus obscurus, 75
Anthus pratensis, 73
Anthus trivialis, 71
Aquila chrysaetus, 197
Archibuteo lagopus, 206
Arctic Tern, 306
Ardea cinerea, 248
Asio brachyotus, 185
Asio otus, 187
Astur palumbarius, 211

Baillon's Crake, 337
Barn Owl, 182
Barn Swallow, 159
Bearded Titmouse, 81

Bittern, 250
Blackbird, 129
Blackcap Warbler, 113
Black Grouse, 360
Black Guillemot, 312
Black-headed Gull, 297
Black-necked Grebe, 333
Black Redstart, 139
Black-throated Diver, 327
Blue-headed Wagtail, 68
Blue Titmouse, 91
Botaurus stellaris, 250
Bullfinch, 28
Buteo vulgaris, 204

Caccabis rufa, 354
Capercaillie, 359
Caprimulgus europæus, 177
Carrion Crow, 3
Certhia familiaris, 77
Chaffinch, 40
Charadrius pluvialis, 255
Chelidon urbica, 161
Chiffchaff, 100
Cinclus aquaticus, 148
Circus æruginosus, 210
Circus cineraceus, 207
Circus cyaneus, 208
Cirl Bunting, 54
Clangula glaucion, 243
Coal Titmouse, 89
Coccothraustes vulgaris, 26
Columba œnas, 346
Columba livia, 348
Columba palumbus, 344
Colymbus arcticus, 327

INDEX

Colymbus septentrionalis, 325
Common Buzzard, 204
Common Chough, 12
Common Coot, 342
Common Creeper, 77
Common Crossbill, 23
Common Curlew, 272
Common Eider, 244
Common Guillemot, 310
Common Gull, 295
Common Heron, 248
Common Jay, 15
Common Kingfisher, 181
Common Kite, 201
Common Nightjar, 177
Common Nuthatch, 79
Common Partridge, 355
Common Quail, 352
Common Sandpiper, 266
Common Scoter, 242
Common Sheldrake, 225
Common Snipe, 282
Common Swift, 175
Common Teal, 232
Common Tern, 304
Common Wren, 150
Cormorant, 216
Corn Bunting, 52
Corn Crake, 334
Corvus corax, 1
Corvus cornix, 5
Corvus corone, 3
Corvus frugilegus, 7
Corvus monedula, 10
Coturnix communis, 352
Cotyle riparia, 163
Crested Titmouse, 85
Crex bailloni, 337
Crex porzana, 335
Crex pratensis, 334
Cuckoo, 172
Cuculus canorus, 172
Cygnus olor, 222
Cypselus apus, 175

Dartford Warbler, 105
Dipper, 148
Dotterel, 257
Dunlin, 278

Emberiza cirlus, 54
Emberiza citrinella, 55
Emberiza miliaria, 52
Emberiza schœniclus, 50
Erithacus luscinia, 135
Erithacus rubecula, 133
Eudromias morinellus, 257

Falco æsalon, 193
Falco peregrinus, 189
Falco subbuteo, 191
Falco tinnunculus, 195
Fratercula arctica, 315
Fringilla carduelis, 36
Fringilla chloris, 34
Fringilla cœlebs, 40
Fringilla spinus, 38
Fulica atra, 342
Fuligula cristata, 240
Fuligula ferina, 239
Fuligula nigra, 242
Fulmar Petrel, 321
Fulmarus glacialis, 321

Gadwall, 227
Gallinula chloropus, 340
Gannet, 220
Garden Warbler, 111
Garganey, 234
Garrulus glandarius, 15
Gecinus viridis, 166
Goldcrest, 95
Golden Eagle, 197
Golden-Eye, 243
Golden Oriole, 21
Golden Plover, 255
Goldfinch, 36
Goosander, 245
Goshawk, 211
Grasshopper Warbler, 122
Gray-Lag Goose, 224
Gray Wagtail, 66
Great Black-backed Gull, 291
Great Crested Grebe, 331
Great Skua, 285
Great Spotted Woodpecker, 170
Great Titmouse, 93
Greater Ringed Plover, 260
Greenfinch, 34

INDEX.

Greenshank, 271
Green Woodpecker, 166

Hæmatopus ostralegus, 262
Haliaëtus albicilla, 199
Hawfinch, 26
Hedge Accentor, 146
Hen Harrier, 208
Herring Gull, 289
Hirundo rustica, 159
Hobby, 191
Honey Buzzard, 203
Hooded Crow, 5
Hoopoe, 179
House Martin, xii, 161
House Sparrow, 29

Iynx torquilla, 164

Jackdaw, 10

Kentish Sand Plover, 259
Kestrel, 195
Kittiwake, 287

Lagopus mutus, 364
Lagopus scoticus, 362
Lanius collurio, 96
Lanius rufus, 99
Lapwing, 253
Larus argentatus, 289
Larus canus, 295
Larus fuscus, 293
Larus marinus, 291
Larus ridibundus, 297
Larus tridactylus, 287
Leach's Fork-tailed Petrel, 317
Lesser Black-backed Gull, 293
Lesser Redpole, 46
Lesser Spotted Woodpecker, 163
Lesser Tern, 308
Lesser Whitethroat, 107
Linnet, 42
Linota cannabina, 42
Linota flavirostris, 44
Linota rufescens, 46
Little Grebe, 329
Locustella locustella, 122
Long-eared Owl, 187

Long-tailed Titmouse, 83
Loxia curvirostra, 23

Magpie, 17
Mallard, 237
Manx Shearwater, 323
Marsh Harrier, 210
Marsh Titmouse, 87
Marsh Warbler, 115
Meadow Pipit, 73
Mergus merganser, 245
Mergus serrator, 247
Merlin, 193
Merula merula, 129
Merula torquata, 131
Milvus regalis, 201
Missel-Thrush, 127
Montagu's Harrier, 207
Motacilla alba, 64
Motacilla alba yarrellii, 62
Motacilla flava, 68
Motacilla raii, 69
Motacilla sulphurea, 66
Muscicapa atricapilla, 157
Muscicapa grisola, 155
Mute Swan, 222

Nightingale, 135
Numenius arquatus, 272
Numenius phæopus, 274

Œdicnemus crepitans, 251
Oriolus galbula, 21
Osprey, 214
Oystercatcher, 262

Pallas's Sand Grouse, 351
Pandion haliaëtus, 214
Panurus biarmicus, 81
Parus ater, 89
Parus ater britannicus, 89
Parus cæruleus, 91
Parus cristatus, 85
Parus major, 93
Parus palustris, 87
Parus palustris dresseri, 87
Passer domesticus, 29
Passer montanus, 32
Perdix cinerea, 355

B B

Peregrine Falcon, 189
Pernis apivorus, 203
Phalacrocorax carbo, 216
Phalacrocorax graculus, 218
Phalaropus hyperboreus, 276
Phasianus colchicus, 35
Pheasant, 357
Phylloscopus rufus, 100
Phylloscopus sibilatrix, 103
Phylloscopus trochilus, 101
Pica caudata, 17
Picus major, 170
Picus minor, 163
Pied Flycatcher, 157
Pied Wagtail, 62
Pintail Duck, 229
Plectrophenax nivalis, 48
Pochard, 239
Podiceps cornutus, 333
Podiceps cristatus, 331
Podiceps minor, 329
Podiceps nigricollis, 333
Pratincola rubetra, 142
Pratincola rubicola, 144
Procellaria leachi, 317
Procellaria pelagica, 319
Ptarmigan, 364
Puffin, 315
Puffinus anglorum, 323
Pyrrhocorax graculus, 12
Pyrrhula vulgaris, 28

Raven, 1
Razorbill, 313
Red-backed Shrike, 96
Red-breasted Merganser, 247
Red Grouse, 362
Red-legged Partridge, 354
Red-necked Phalarope, 276
Redshank, 269
Redstart, 137
Red-throated Diver, 325
Reed Bunting, 50
Reed Warbler, 117
Regulus cristatus, 95
Richardson's Skua, 283
Ring Dove, 344
Ring Ouzel, 131
Robin, 133

Rock Dove, 348
Rock Pipit, 75
Rook, 7
Roseate Tern, 302
Rough-legged Buzzard, 206
Ruff, 264
Ruticilla phœnicurus, 137
Ruticilla tithys, 139

St. Kilda Wren, 152
Sand Martin, 163
Sandwich Tern, 300
Saxicola œnanthe, 140
Sclavonian Grebe, 333
Scolopax gallinago, 282
Scolopax rusticola, 280
Sedge Warbler, 120
Shag, 218
Short-eared Owl, 185
Shoveller, 235
Siskin, 38
Sitta cæsia, 79
Sky-Lark, 58
Snow Bunting, 48
Somateria mollissima, 244
Song Thrush, 125
Sparrow-Hawk, 212
Spotted Crake, 335
Spotted Flycatcher, 155
Starling, 19
Stercorarius catarrhactes, 285
Stercorarius richardsoni, 283
Sterna arctica, 306
Sterna cantiaca, 300
Sterna dougalli, 302
Sterna hirundo, 304
Sterna minuta, 308
Stock Dove, 346
Stonechat, 144
Stone Curlew, 251
Stormy Petrel, 319
Strix aluco, 184
Sturnus vulgaris, 19
Sula bassana, 220
Sylvia atricapilla, 113
Sylvia cinerea, 109
Sylvia curruca, 107
Sylvia hortensis, 111
Sylvia provincialis, 105

INDEX.

Syrrhaptes paradoxus, 351

Tadorna cornuta, 225
Tetrao tetrix, 360
Tetrao urogallus, 359
Totanus calidris, 269
Totanus glareola, 267
Totanus glottis, 271
Totanus hypoleucus, 266
Totanus pugnax, 264
Tree Pipit, 71
Tree Sparrow, 32
Tringa alpina, 278
Troglodytes parvulus, 150
Troglodytes parvulus hirtensis, 152
Tufted Duck, 240
Turdus musicus, 125
Turdus viscivorus, 127
Turtle Dove, 350
Turtur auritus, 350
Twite, 44

Upupa epops, 179
Uria grylle, 312

Uria troile, 310

Vanellus cristatus, 253

Waterhen, 340
Water-Rail, 338
Wheatear, 140
Whimbrel, 274
Whinchat, 142
White-tailed Eagle, 199
Whitethroat, 109
White Wagtail, 64
Wigeon, 230
Willow Wren, 101
Woodchat Shrike, 99
Woodcock, 280
Wood-Lark, 60
Wood Owl, 184
Wood Sandpiper, 267
Wood Wren, 103
Wryneck, 164

Yellow Bunting, 55
Yellow Wagtail, 69

RICHARD CLAY & SONS, LIMITED,
LONDON & BUNGAY.

CHARLES DIXON'S WORKS.

THE NESTS AND EGGS OF NON-INDIGENOUS BRITISH BIRDS.

Crown 8vo.

JOTTINGS ABOUT BIRDS.

With Coloured Frontispiece by J. SMIT. Crown 8vo, 6s.

The Times says:—"A very pleasant series of papers on bird life and bird distribution by the well known author of 'The Migration of Birds,' and other works on ornithology. Mr. Dixon's observation is close and his knowledge of birds is accurate and extensive. In one paper he makes good fun of the mistakes which artists often make in depicting birds; in another he discusses the cuckoo and its habits; in another he propounds a theory of the purely amatory origin and character of the song of birds; while papers on the Bass Rock and on St. Kilda are full of charming description. All lovers of birds will read the book with delight."

The Scotsman says:—"Is pleasantly discursive, teems with useful and interesting information, and forms agreeable reading, whether for the initiated ornithologist, or for the reader who, without much special knowledge, has a genuine love for the woodlands and their feathered population. . . . In this volume Mr. Dixon takes us on a charming tour to Algeria, and gives the best account yet available to English readers of the birds of that country. . . . A very interesting book."

THE NESTS AND EGGS OF BRITISH BIRDS:

WHEN AND WHERE TO FIND THEM.

Being a Handbook to the Oology of the British Islands.

Crown 8vo, 6s.

Natural Science says:—"Like all Mr. Dixon's productions, the work is well and pleasantly written, and the amateur naturalist, as well as every young person interested in this fascinating study, cannot do better than forthwith provide him- or herself with a copy . . . . fully deserves all the success we can wish it."

The Speaker says:—"Mr. Charles Dixon has written an admirable manual—the outcome of the personal observation and patient research of many years spent amid rural surroundings. . . . We know of no other book of similar compass which gives more explicit information concerning the habitat of British birds, their breeding habits, &c. . . . . Everywhere the book displays close acquaintance with the characteristics of wild life in the woods and hedgerows."

THE GAME BIRDS AND WILD FOWLS OF THE BRITISH ISLANDS.

Being a Handbook for the Naturalist and Sportsman.

Illustrated by A. T. ELWES. Demy 8vo, 18s.

The Times says:—"All sportsmen and naturalists and all who love birds, even if they do not claim to be naturalists, and do not even desire to be sportsmen, will welcome Mr. Charles Dixon's elaborate and comprehensive work on 'The Game Birds and Wild Fowl of the British Islands.' Mr. Dixon is the author of the work on 'The Migration of Birds,' which we noticed not long ago, and of many other books relating to bird-life and the study of rural nature. Mr. Dixon's present work is full of interest for the bird-lover, and full of information for the sportsman, besides being copious and exact from the purely scientific point of view."

The Daily Telegraph says:—"A valuable work. . . . The information which is given in the case of each family of birds is comprehensive; it includes a description of habits and appearance, which is often supplemented by excellent illustrations in black and white, an account of its geographical distribution, lines of migration, and manner of building its nest. Where personal experience has fallen short, the author has gone to the highest and latest authorities; and the result is a book which will be of great assistance to both the classes for whom it is intended to cater."

CHAPMAN & HALL, LIMITED, LONDON.

CHARLES DIXON'S WORKS (continued).

THE MIGRATION OF BIRDS.
An Attempt to Reduce Avian Season-Flight to Law.
Crown 8vo, 6s.

The Times says:—"Mr. Charles Dixon, than whom, perhaps, no more scientific ornithologist exists, formulates a theory to account for a phenomenon which has hitherto refused to yield up its secret. He first dismisses rather contemptuously the view of those naturalists who ascribe migration to instinct. For 'instinct' he would substitute 'habit.' The superiority of Mr. Dixon's theory really resides in this, that he offers a rational explanation of the origin of this 'instinct' or 'hereditary impulse.' . . . The plausible theory which Mr. Dixon propounds is illustrated with abundance of ornithological learning, and a multitude of examples which, he tells us, might have been indefinitely increased. Apart from his speculations, Mr. Dixon's book is a most interesting monograph upon the facts and phenomena of bird migration, and we can hardly doubt that, whether his theories win acceptance or not, the volume in which he sets them forth will become part of the necessary equipment for future explorers in this department of ornithology."

Dr. Andrew Wilson says:—"Among recently published scientific books, there are two which I think worth recommending to the notice of my readers. The first of these works is one on 'The Migration of Birds,' by C. Dixon. It deals in an exhaustive manner with migration at large, and cannot fail to interest all who, in any fashion, make ornithology a study."

THE BIRDS OF OUR RAMBLES:
A Companion for the Country.
With Illustrations by A. T. ELWES. Crown 8vo, 7s. 6d.

The Globe says:—"In 'The Birds of our Rambles' we have yet another of Mr. Charles Dixon's popular descriptions of natural objects. . . . His object is less to be severely scientific than to be pleasantly graphic, his method being to direct the observer's attention to whatever, in the birds mentioned, is most likely to appeal to him—the notes, the general appearance, or any peculiarity of habit. The result is a book which though practically encyclopædic in comprehensiveness and detail, is nevertheless eminently readable. Some excellent illustrations help to assist the text."

IDLE HOURS WITH NATURE.
With Frontispiece. Crown 8vo, 6s.

Black and White says:—"The title of Mr. Charles Dixon's 'Idle Hours with Nature' is a somewhat exasperating misnomer. So far from being idle, he is one of the busiest observers of nature since White of Selborne wrote, or the modern White, Richard Jefferies. . . . George Eliot used to say that anglers could not catch fish because they would not study the *subjectivity* of fishes. Mr. Dixon studies the subjectivity of the wild birds and beasts in a way that has never been done before, and his book is profoundly interesting in consequence. He enters into the minds and moods of the creatures of the air, large and small, and reasons from his observations. He tells us what the Spotted Fly-Catcher must see and feel and desire as it flits on its long migration from the Sahara to its home in our English apple orchards; and he analyzes the character of cormorants, petrels, and eagles as carefully and conclusively as a novelist does his heroes and his villains. Mr. Dixon contends that his studies have an ethical value beyond their scientific one. There can be no doubt about it—they take us out of ourselves."

ANNALS OF BIRD LIFE:
A Year-Book of British Ornithology.
With Illustrations by C. WHYMPER. Crown 8vo, 7s. 6d.

The Speaker says:—"Delightful book . . . . In this volume five or six chapters are devoted, in turn, to spring, summer, autumn, and winter; and everywhere, without thrusting upon us the dry details of science or the jargon of the schools, a minute and pleasing description is given of the way of birds, their migration, and the gipsy kind of life they lead."

The Leeds Mercury says:—"Full of restful charm of rural life written with considerable ability and a real enthusiasm for the subject. The work is the outcome of twenty years' close study and observation of wild life in woods and fields, and beginning with spring, it takes the reader right through the year, and shows him at each season the various movements and habits of the birds. . . . The book is a fresh, artless, and minute description of Nature at first hand."

CHAPMAN & HALL, LIMITED, LONDON.

W. H. HUDSON'S WORKS.

BIRDS IN A VILLAGE.
Square Crown 8vo, 7s. 6d.

The Academy says:—"Mr. Hudson is a very loving student of birds. No movement, no twitter, no cadence of song escapes him; and his analytic mind at once asks the reason of all these changeful habits. . . . Mr. Hudson's style is admirable: at the same time lucid and attractive. . . . No rules, no mannerism bind Mr. Hudson. His keen and subtle powers of observation are seconded by a playful fancy; while a rich imaginative halo is thrown round the bird he describes, which brings it into greater prominence, as it were, and strongly impresses its individuality upon the reader. . . . In short, this whole book is delightful, and any kind of praise or commendation is superfluous."

The Graphic says:—"The book is one to read, delight in, and appreciate. Mr. Hudson belongs to the school of Thoreau and of John Burroughs; but he is better than the last because he is more suggestive, with a finer vein of poetry running like a fugue through his writing."

THE NATURALIST IN LA PLATA.
With Illustrations by J. SMIT. Demy 8vo, 16s.

Mr. Alfred R. Wallace in "Nature" says:—"This volume, so far as the present writer knows, is altogether unique among books on natural history. What renders this work of such extreme value and interest is, that it is not written by a traveller or a mere temporary resident, but by one born in the country, to whom its various tribes of beasts, birds, and insects have been familiar from childhood; who is imbued with love and admiration for every form of life; and who for twenty years has observed carefully and recorded accurately everything of interest in the life-histories of the various species with which he has become acquainted. The book is written in an earnest spirit, and in a clear and delightful style. . . . It remains only to add that the book is beautifully got up, and that the numerous illustrations are at once delicate and characteristic. Never has the present writer derived so much pleasure and instruction from a book on the habits and instincts of animals, the most interesting and delightful of modern books on natural history."

IDLE DAYS IN PATAGONIA.
Illustrated by ALFRED HARTLEY and J. SMIT. Demy 8vo, 14s.

The Times says:—"'Idle Days in Patagonia' is a welcome and worthy addition to the literature of travel and zoological observations in South America—already so rich by the labours and writings of Bates, Darwin, and of Mr. Hudson himself, who is not unworthy to be named in this distinguished company. Mr. Hudson is a keen observer, an acute reasoner, and a very attractive writer, and the many readers who have appreciated his 'Naturalist in La Plata' will turn with eagerness to his 'Idle Days in Patagonia,' and will not be disappointed."

The Scotsman says:—"In the new volume Mr. Hudson presents himself almost more as the poet than as the observer of wild nature. . . . Mr. Hudson's chief field of Patagonian research was on the Rio Negro, whose valley is a strip of life and greenness drawn through the dry and thorny wilderness. There is not a dull or an unsuggestive page in his book. Personal adventures there are not a few, but Mr. Hudson is almost more interesting when he turns aside to meditate upon, and illustrate from the rich store of his reading and experience, such themes as bird music, migratory instincts, the 'quality of whiteness' in snow and other natural objects, the mysteries enfolded in the sense of smell, keenness of sight and colour sense in savage and in civilized men, and the predominating colour of the eye, and its significance in different races of mankind."

CHAPMAN & HALL, LIMITED, LONDON.

EXTINCT MONSTERS.
A Popular Account of some of the Larger Forms of Ancient Animal Life.
By the Rev. H. N. HUTCHINSON.
With numerous Illustrations by J. SMIT and others, and a Preface by DR. HENRY WOODWARD, F.R.S. Demy 8vo, 12s.

SOME OPINIONS OF THE PRESS.

The Geological Magazine says:—"The author indulges in no rodomontade, but gives a careful and readable account of the wonderful discoveries which modern geological research in the Rocky Mountains, and in many other parts of the world, has brought to light. The twenty-four full-page illustrations are admirably executed, and there are thirty-eight others in the text. There is a freshness about the whole thing which suggests 'Alice in Wonderland.' The book is a safe book to put into the hands of the young, and cannot fail to interest geologists of all ages. It will make an admirable and attractive New Year's book, which every one should buy and read for themselves."

The Saturday Review says:—"Mr. Hutchinson writes pleasantly and unaffectedly, combining much information of scientific value with many interesting anecdotes of the discovery of fossils, and the legends which have gathered round them. His book, in short, is both attractive and useful, and will add to his reputation as a popular, but accurate, writer on geological subjects."

The Athenæum says:—"This is undoubtedly the best book Mr. Hutchinson has yet written. He sets before us, in pleasant form, a really valuable description of many of those extraordinary forms of ancient life which are but little known, save to the special student of palæontology. Not content with the dry bones which have been unearthed by the spade and pick of the geological explorer, the author seeks to revivify these relics, and to place them before us as they probably appeared when clothed with flesh and instinct with life. Mr. Hutchinson has been fortunate in receiving the advice of the Geological Department of the British Museum and the skilful pen of Mr. Smit, who is probably unsurpassed as a scientific artist of animals."

The Field says:—"In 'Extinct Monsters' the Rev. H. N. Hutchinson has admirably succeeded in his aim at giving a popular account of the larger forms of animal life. In the present volume we have a competent palæontologist, who has availed himself of the most recent discoveries, and has been assisted by some of the first geologists in the kingdom, who have taken a great interest in the work, and whose revision of the proof-sheets gives us a guarantee that, astounding as the statements of Mr. Hutchinson appear, and monstrous as the illustrations of Mr. Smit undoubtedly are, they describe and actually represent the extinct forms which have recently come to light. It is thoroughly readable."

Black and White says:—"M. Cuvier's vast and splendid knowledge of existing beasts and birds enabled him to reconstruct from a fossil skull or a vertebra, sometimes from nothing but a single tooth, the long extinct creature in its true semblance as it had lived—to clothe it with flesh and skin and show it, in imagination, in the haunts in which it lived and moved. This, which Baron Cuvier did in graphic description of great scientific and literary beauty, Mr. Hutchinson, in his work on 'Extinct Monsters,' has done popularly and done learnedly, and with the accompaniment of many most admirable illustrations. . . . This learned, interesting, and popular book."

By the same Author.
CREATURES OF OTHER DAYS.
By REV. H. N. HUTCHINSON, F.G.S., Author of 'Extinct Monsters.'
With Illustrations by J. SMIT. Large Crown 8vo.

CHAPMAN & HALL, LIMITED, LONDON.

www.ingramcontent.com/pod-product-compliance
Lightning Source LLC
Chambersburg PA
CBHW020535300426
44111CB00008B/681